Foreword

From the Director
U.S. Army Capabilities Integration Center

I0468628

The U.S. Army Training and Doctrine Command's assessment of the future operational environment highlights the importance of all aspects of information on the future battlefield. Army forces operate in and among human populations, facing hybrid threats that are innovative, networked, and technologically-savvy. These threats capitalize on emerging technologies to establish and maintain a cultural and social advantage; leveraging these new capabilities for command and control, recruiting, coordinating logistics, raising funds, and propagandizing their message. To operate effectively in this emerging environment, the Army must realign its information "Aim Point." Army leaders and Soldiers must possess an in-depth understanding of how to leverage information-based capabilities to gain and maintain situational awareness. Understanding how to fight for and leverage the power of information, while denying the adversary's ability to do the same, will be increasingly critical to success on the future battlefield.

The assessment indicates that the Army's current vocabulary, including terms such as computer network operations (CNO), electronic warfare (EW), and information operations (IO) will become increasingly inadequate. To address these challenges, there are three interrelated dimensions of full spectrum operations (FSO), each with its own set of causal logic, and requiring focused development of solutions:

- The first dimension is the psychological contest of wills against implacable foes, warring factions, criminal groups, and potential adversaries.
- The second dimension is strategic engagement, which involves keeping friends at home, gaining allies abroad, and generating support or empathy for the mission.
- The third dimension is the cyber-electromagnetic contest, which involves gaining, maintaining, and exploiting a technological advantage.

The first and second dimensions focus on how commanders and staffs orchestrate and leverage information power to achieve their missions. The third dimension focuses on gaining and maintaining an advantage in the converging mediums of cyberspace and the electromagnetic spectrum (EMS). The Army's construct of gaining advantage, protecting that advantage, and placing adversaries at a disadvantage is well nested within these dimensions; and contributes to the outcomes that must be achieved by unified action at the tactical, operational, and strategic levels. Current operations reinforce our conviction that concepts and capabilities are needed for each of these dimensions.

IO encompasses all three of these dimensions, but is increasingly an overburdened term which refers to any use of information. CNO and EW by themselves are insufficient to describe the full scope of the cyber-electromagnetic contest. To this end, we are undertaking a comprehensive campaign to describe fully each dimension. The first two dimensions (the contest of wills and strategic engagement), will be addressed in a forthcoming, separate concept capability plan, and followed by a capability based assessment.

This pamphlet relates EW, CNO, and cyber in this third dimension–the cyber-electromagnetic contest. TRADOC Pam 525-7-8 is the first step in developing a common understanding of how technological advancements transform the operational environment, how leaders must think about cyberspace operations, how they should integrate their overall operations, and which capabilities are needed. It provides the means to identify outcomes-based, integration-focused, and resource-informed solutions which enable the U.S. Army to prevail in the cyber-electromagnetic contest.

MICHAEL A. VANE
Lieutenant General, U.S. Army
Director, Army Capabilities
Integration Center

Executive Summary

Framing the problem
The operational environment (OE) has changed dramatically. The technologic convergence of computer and telecommunication networks; astonishing rates of technologic advancements; global proliferation of information and communications technology (ICT) and its consequent effect in social networks and in society impact the OE. The diverse and wide arrays of agents who use or exploit this technological revolution pose a grave threat to U.S. critical infrastructure and operational missions. These agents range from traditional nation-states to noncombatants, transnational corporations, criminal organizations, terrorists, hacker unions, mischievous hackers, and the unwitting individual who intends no malice. Collectively, they combine to create a condition of perpetual turbulence without traditional end states or resolution. Unless otherwise noted in this document, the terms "adversary" and "adversaries" are used in this broad context.

Framing the solution
Training and Doctrine Command (TRADOC) Pamphlet (Pam) 525-7-8, *The U.S. Army Concept Capability Plan for Cyberspace Operation* (CyberOps) *2016-2028,* takes a comprehensive look at how the Army's future force in 2016-2028 will leverage cyberspace and CyberOps. This pamphlet includes a conceptual framework for integrating CyberOps into FSO, thereby providing the basis for follow-on doctrine development efforts. This conceptual framework outlines how commanders integrate CyberOps to gain advantage, protect that advantage, and place adversaries at a disadvantage. This pamphlet also establishes a common lexicon for Army CyberOps, and describes the relationship between cyberspace, the other four domains (air, land, maritime, and space), and the EMS. Lastly, it explains how converging technologies will increasingly affect FSO and influence capability development, thereby enabling the Army to influence the design, development, acquisition, and employment of fully integrated cyber capabilities.

Solution context: the three dimensions of FSO

a. The Commanding General (CG), U.S. Army TRADOC directed the Combined Arms Center (CAC) to lead a working group to establish the conceptual framework for the organization for the cyberspace (cyber), EW, and IO mission areas and TRADOC's associated force modernization proponency structure. On 16 October 2009, the CG TRADOC provided recommendations to the Army, Vice Chief of Staff. Included among his recommendations were the following:

(1) The CAC determined that current vocabulary (cyber-EW-IO) is adequate today, but will become increasingly inadequate to describe the challenges the Army faces in the operational environment.

(2) The CAC concluded that there are three dimensions to be addressed, that these dimensions exist across the FSO, and that these dimensions each require force design and doctrinal solutions.

(3) Therefore, although the Army currently describes the functions related to network and spectrum operations as cyber-EW-IO, the CAC believe that the Army should adapt and describe them in the future as follows:

- First dimension - The first dimension is the psychological **contest of wills** against implacable foes, warring factions, criminal groups, and potential adversaries.
- Second dimension - The second dimension is **strategic engagement** and involves keeping friends at home, gaining allies abroad, and generating support or empathy for the mission in the area of operations.
- Third dimension - The third dimension is the **cyber-electromagnetic contest**[1]. Trends in wired, wireless, and optical technologies are setting conditions for the convergence of computer and telecommunication networks.

b. TRADOC Pam 525-7-8 is fundamentally about prevailing in this third dimension, the cyber-electromagnetic contest, and provides recognition that CyberOps enables the first two dimensions.

Central idea

a. Prevailing in the cyber-electromagnetic contest means making progress at the same time along three lines of effort: gaining advantage, protecting that advantage, and placing adversaries at a disadvantage.

b. Commanders seek to retain freedom of action in cyberspace and in the EMS, while denying the same to adversaries at the time and place of their choosing; thereby enabling operational activities in and through cyberspace and consequently the other four domains. CyberOps encompass those actions to gain the advantage, protect that advantage, and place adversaries at a disadvantage in the cyber-electromagnetic contest. CyberOps are not an end to themselves, but rather an integral part of FSO and include activities prevalent in peacetime military engagement, which focus on winning the cyber-electromagnetic contest. CyberOps are continuous; engagements occur daily, most often without the commitment of additional forces.

Solution framework
Current doctrinal terms do not adequately address the broad range of tasks associated within Department of Defense (DOD) definitions of cyberspace and CyberOps. Consequently, the framework developed for TRADOC Pam 525-7-8 establishes four components for CyberOps: cyber warfare (CyberWar), cyber network operations (CyNetOps), cyber support (CyberSpt) and cyber situational awareness (CyberSA).

---

[1] The use of the term cyber-electromagnetic is not meant to equate the terms cyberspace and electromagnetic spectrum, but rather to highlight there is significant overlap between the two and future technological development is likely to increase this convergence.

Department of the Army
Headquarters, United States Army
Training and Doctrine Command
Fort Monroe, Virginia 23651-1047

TRADOC Pamphlet 525-7-8

22 February 2010

Military Operations
CYBERSPACE OPERATIONS CONCEPT CAPABILITY PLAN 2016-2028

FOR THE COMMANDER:
OFFICIAL:

DAVID P. VALCOURT
Lieutenant General, U.S. Army
Deputy Commanding General/
 Chief of Staff

LUCIOUS B. MORTON
Colonel, GS
Deputy Chief of Staff, G-6

History. This publication is a new U.S. Army Training and Doctrine Command (TRADOC) concept capability plan developed as part of the Army Concept Framework for the future force and as part of the capabilities-based assessment (CBA) process.

Summary. TRADOC Pam 525-7-8 takes a comprehensive look at how the Army's future force in 2016-2028 will leverage cyberspace and CyberOps. This pamphlet includes a conceptual framework for integrating CyberOps into full spectrum operations (FSO), thereby providing the basis for follow-on doctrine development efforts. This conceptual framework outlines how commanders integrate CyberOps to gain advantage, protect that advantage, and place adversaries at a disadvantage. This pamphlet establishes a common lexicon for Army CyberOps, and describes the relationship between cyberspace, the other four domains (air, land, maritime, and space), and the electromagnetic spectrum (EMS). TRADOC Pam 525-7-8 explains how converging technologies will increasingly affect FSO and influence capability development; thereby enabling the Army to influence the design, development, acquisition, and employment of fully integrated cyber capabilities.

Applicability. TRADOC Pam 525-7-8 is the foundation for future force development and the base for subsequent developments of supporting concepts, concept capability plans, and the Joint Capabilities Integration and Development System (JCIDS) process. It supports experimentation described in the Army Capabilities Integration Center (ARCIC) Campaign Plan and functions as the basis for developing solutions related to the future force within the doctrine, organizations, training, materiel, leadership and education, personnel, and facilities (DOTMLPF) domains. This pamphlet applies to all TRADOC, Department of Army (DA) and Army Reserve component activities that develop DOTMLPF requirements.

Proponent and supplementation authority. The proponent of this pamphlet is the TRADOC Headquarters, Director, ARCIC. The proponent has the authority to approve exceptions or waivers to this pamphlet that are consistent with controlling law and regulations. Do not supplement this pamphlet without prior approval from Director, TRADOC ARCIC (ATFC-ED), 33 Ingalls Road, Fort Monroe, VA 23651-1061.

Suggested Improvements. Users are invited to submit comments and suggested improvements via The Army Suggestion Program online at https://armysuggestions.army.mil (Army Knowledge Online account required) or via DA Form 2028 to Director, TRADOC ARCIC (ATFC-ED), 33 Ingalls Road, Fort Monroe, VA 23651-1061. Suggested improvements may also be submitted using DA Form 1045.

Availability. This regulation is available on the TRADOC homepage at http://www.tradoc.army.mil/tpubs/regndx.htm

Contents   **Page**

Table List

Figure List

Chapter 1
Introduction

1-1. Relevance

a. The operational environment (OE) has changed dramatically. Unprecedented levels of adverse activity in and through cyberspace threaten the integrity of United States (U.S.) critical infrastructure, financial systems, and elements of national power. These threats range from unwitting hackers to nation-states, each at various levels of competence. Collectively, the threats create a condition of perpetual turbulence without traditional end states or resolution. Unless otherwise noted in this document, the terms "adversary" and "adversaries" are used in this broad context.

b. The ever-increasing rate of technologic advances and its wide proliferation make it increasingly difficult to achieve success across the military FSO. The convergence of wired, wireless, and optical technologies has led to the merging of computer and telecommunication networks; handheld computing devices continue to grow in number and capability. Next generation systems are beginning to emerge, forming a global, hybrid, and adaptive network that combines wired, wireless, optical, satellite communications, supervisory control, and data acquisition (SCADA), and other systems. Soon networks will provide ubiquitous access to users and enable them to collaborate when needed in near real time.

c. The Nation's adversaries' ability to stay apace with the accelerating rate of technologic change complicates the OE. A significant advantage will go to the side that gains, protects, and exploits advantage in the contested and congested cyberspace and EMS. Conversely, the side that fails in this contest, or that cannot operate effectively when their systems are degraded or disrupted, cedes a significant advantage to the adversary.

d. Gaining, protecting, and exploiting the advantage will not be easy. U.S. adversaries use the commercial marketplace as their combat developer, which makes them much more nimble and adaptive than the Army's lengthy research, development, test, evaluation, and acquisition processes. Adversaries increasingly capitalize on cyberspace and electromagnetic capabilities and activities, while to date those capabilities and activities too often have been peripheral to our Army's normal operations. To seize and maintain the operational and tactical advantage against such adaptive adversaries, Army forces must make cyberspace and the EMS central and routine components of its operations; and commanders will need, among other things, the associated capabilities, and the corresponding subject matter expertise to apply them.

1-2. Purpose
The purpose of TRADOC Pam 525-7-8 is to examine how the Army's future force in 2016-2028 will integrate cyberspace capabilities and CyberOps as part of FSO.

1-3. Scope
TRADOC Pam 525-7-8 provides an initial examination of how CyberOps are integrated with the commander's other capabilities to gain advantage, to protect that advantage, and to place adversaries at a disadvantage in FSO. The examination will be refined through the CBA and

doctrine development process. This pamphlet describes how commanders seek to retain freedom of action in cyberspace and in the EMS, while denying the same to their adversaries at the time and place of the Army's choosing; thereby enabling other operational activities in and through cyberspace as well as in the other four domains. This pamphlet establishes a common lexicon and framework for CyberOps and describes the relationship between cyberspace, the air, land, maritime and space domains, and the EMS. It also explains how converging technologies will increasingly affect FSO and influence capability development; identifies CyberOps and enabling capabilities needed to support future force modernization initiatives; and presents cyberspace and EMS study issues suitable for experimentation.

1-4. Method

This pamphlet leverages the TRADOC-approved design process. Chapter 2 describes the existing and desired conditions of the operational environment as they pertain to cyberspace. Chapter 3 compares the existing conditions in the operational environment to the desired end state; thereby establishing the hypothesis for framing the solution. Chapter 4 establishes the framework, central and supporting ideas, and lexicon. Appendix A contains the required and related references. Appendix B introduces the evolving cyber operational structure. Appendix C describes how CyberOps are integrated as part of the overall operation to achieve the commander's intent and objectives, and not an end to themselves. Appendices D (unclassified) and E (classified) discuss required capabilities. Appendix F provides the operative questions across DOTMLPF to help with the initial steps of the ensuing CBA.

1-5. Key definitions

a. Cyberspace is defined as, "A global domain within the information environment consisting of the interdependent network of information technology infrastructures, including the Internet, telecommunications networks, computer systems, and embedded processors and controllers."[2]

b. CyberOps are, "The employment of cyber capabilities where the primary purpose is to achieve objectives in or through cyberspace. Such operations include computer network operations and activities to operate and defend the global information grid (GIG)."[3]

c. EMS is the range of frequencies of electromagnetic radiation from zero to infinity. It is divided into 26 alphabetically designated bands."

1-6. Relation to joint and Army concepts

a. TRADOC Pam 525-7-8 is compatible with joint and Army concepts including the Capstone Concept for Joint Operations and the Army capstone concept. The capabilities described in this pamphlet are nested with the joint capability areas (JCA) and warfighting

---

[2] Deputy Secretary of Defense Memorandum, dated 12 May 2008, defined cyberspace. This pamphlet is anchored in the approved DOD definition of cyberspace but there are still multiple perspectives as to the characterization of cyberspace as a domain.

[3] Deputy Secretary of Defense Memorandum, dated 15 October 2008, defined CyberOps. The memo also states that operations which may cause effects in cyberspace (such as, EW, psychological operations) but do not employ cyber capabilities should not be considered CyberOps; and it recommends the common usage of the modifier "cyber" to mean "cyberspace" (such as, cyber attack, cyber defense, and CyberOps.) A Chief, Joint Chief of Staff memo, dated 18 August 2009, updated the DOD definition for cyberspace operations.

functions. The DOD uses JCAs to describe how capabilities support the joint functions. JCAs form the basis of the DOD's capabilities based processes and CyberOps capabilities are nested under the Tier 1 JCAs of force application, protection, battlespace awareness, and net-centric operations. In the same way, CyberOps capabilities enable and are an integral part of the Army's warfighting functions and elements of combat power.

b. TRADOC Pam 525-3-0. TRADOC Pam 525-3-0 recognizes that war is a contest of wills and in order to prevail, the Army must exert a psychological and technical influence as one of the concept's six supporting ideas. The capstone concept states that Army forces are increasingly dependent on electromagnetic, computer network, and space-based capabilities that are converging; therefore exerting technical influence will require forces that are prepared to fight and win on an emerging "cyber-electromagnetic battleground." Because technology that effects how information moves changes so rapidly, the Army must evaluate continuously what competencies and capabilities are required to gain, protect, and exploit advantages in highly contested cyberspace and EMS. This pamphlet supports the capstone concept by identifying required capabilities necessary for successful FSO.

c. TRADOC Pam 525-3-1 and TRADOC Pam 525-3-2. These pamphlets support the Army's operating concepts by identifying the required capabilities for battle command, intelligence, fires, and protection required to execute effective operational and tactical maneuver in the future operational environment. Cyber capabilities and leveraging cyberspace are critical for the Army's future force to be able to command and control on the move while reducing operational risk. TRADOC Pam 525-7-8 reinforces the Army functional concepts that support operational and tactical maneuver.

d. TRADOC Pam 525-7-6. With the convergence of wired, wireless, and optical technologies, the future force commander will use EW and CyberOps capabilities in combination. The increased usage of wired and optical technologies will require that these forces have unimpeded access to the EMS and at the same time be able to deny the adversary use of the same. TRADOC Pam 525-7-6 explores current and required future EW capabilities necessary to maintain the requisite access to the EMS.

e. TRADOC Pam 525-7-16. EMSO capabilities, policies, and coordination are critical for CyberOps activities because of the increased use of wireless technologies. EMSO aims to ensure that electronic systems relying on wireless connectivity are able to perform their functions when and where necessary without causing or suffering interference.

f. TRADOC Pam 525-7-4. Space capabilities enable, and may be enabled by the conduct of CyberOps. Space capabilities are employed in the extension of, and as another transport mode for, the Army's portion of the GIG, LandWarNet, particularly in support of deployed forces. TRADOC Pamphlet 525-7-4 describes the relationship between the space and cyberspace domains.

g. TRADOC Pamphlet 525-5-600. This pamphlet builds on the LandWarNet Concept of Operations' (CONOPS') description of how the Army interfaces with the joint force GIG and conducts network operations (NetOps). This CONOPS details how cyber NetOps is a fundamental element of CyberOps.

1-7. References
Required and related publications are listed in appendix A.

1-8. Explanation of abbreviations and terms
Abbreviations and special terms used in this pamphlet are explained in the glossary.

Chapter 2
Framing the Environment

2-1. Cyberspace, the domain

a. This chapter describes the existing and desired cyberspace conditions of the operational environment in order to enable the framing of the problem in the subsequent chapter. Two authoritative sources serve as the base documents for this chapter: The Joint Forces Command, *Joint Operating Environment 2008: Challenges and Implications for the Future Joint Force*, and the U.S. Army TRADOC *Operational Environment 2009-2025*.

b. Cyberspace is one of five domains; the others are air, land, maritime, and space. These five domains are interdependent. Cyberspace nodes physically reside in all domains. Activities in cyberspace can enable freedom of action for activities in the other domains, and activities in the other domains can also create effects in and through cyberspace.

c. Cyberspace can be viewed as three layers (physical, logical, and social) made up of five components (geographic, physical network, logical network, cyber persona, and persona) (see figure 2-1).

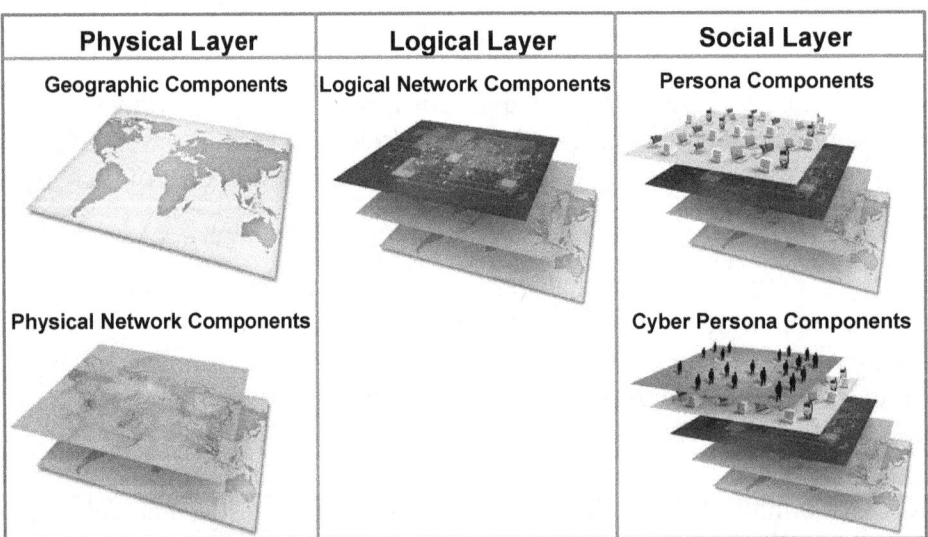

Figure 2-1. The three layers of cyberspace

(1)   The physical layer includes the *geographic component* and the *physical network component*. The geographic component is the physical location of elements of the network. While geopolitical boundaries can easily be crossed in cyberspace at a rate approaching the speed of light, there is still a physical aspect tied to the other domains. The physical network component includes all the hardware and infrastructure (wired, wireless, and optical) that supports the network and the physical connectors (wires, cables, radio frequency, routers, servers, and computers).

(2)   The logical layer contains the *logical network component* which is technical in nature and consists of the logical connections that exist between network nodes. Nodes are any devices connected to a computer network. Nodes can be computers, personal digital assistants, cell phones, or various other network appliances. On an Internet protocol (IP) network, a node is any device with an IP address.

(3)   The social layer comprises the human and cognitive aspects and includes the *cyber persona component* and the *persona component*. The cyber persona component includes a person's identification or persona on the network (e-mail address, computer IP address, cell phone number, and others). The persona component consists of the people actually on the network. An individual can have multiple cyber personas (for example, different e-mail accounts on different computers) and a single cyber persona can have multiple users (for example, multiple users accessing a single eBay® account). This holds important implications for Army forces in terms of attributing responsibility and targeting the source of cyber action. It also means Army forces will require significant situational awareness (SA), forensic, and intelligence capabilities to counter the complex cyber threat.

d.   Cyberspace consists of many different nodes and networks. Though not all nodes and networks are globally connected or accessible, cyberspace continues to become increasingly interconnected. It is easy to traverse geographic boundaries using the Internet when compared to other transmission or travel mediums. Networks, however, can be isolated using protocols, firewalls, encryption, and physical separation from other networks and are typically grouped into domains such as .mil, .gov, .com, and .org. These domains are specific to an organization or mission and organized by physical proximity or function. While some access is achieved globally or remotely, access to closed and specialized networks may require physical proximity.

e.   Advancements in wireless and optical technologies have led to the convergence of computer and telecommunications networks that are increasingly reliant upon portions of the EMS. As technology advances, competition over this portion of the EMS will increase. EW and CyberOps will both require increasing access to the EMS for effective operations.

2-2. Cyberspace and the OE

a.   As stated in FM 3-0, the OE includes physical areas, the information that shapes it, and enemy, adversary, friendly, and neutral systems relevant to a particular operation. This core Army doctrine emphasizes the use of eight interrelated operational variables (political, military, economic, social, information, infrastructure, the physical environment, and time) to understand

and analyze the unique environment in which the Army is conducting operations. Cyberspace and CyberOps are inexorably linked to each of these operational variables.

b. The OE contains unprecedented amounts of information transmitted over commercial networks. As a society, citizens rely on this information and on cyberspace for financial, business, communications, social, and other aspects for daily living. In fact, the 2003 National Strategy to Secure Cyberspace asserts, "cyberspace is our nervous system – our country's control system." According to the 2008 Joint Operating Environment, the global commons have expanded to include the cyberspace domain. [4] This capability is increasingly penetrating less developed areas, enabling more and more populations to gain greater access to these networks.

c. Technology is evolving at astonishing rates and the proliferation of information and communications technology (ICT)[5] has changed the context in which governments and militaries operate. From the 24-hour news cycle, to flash mobs, blogs, social networking, and text messaging, the rapid flow of information has changed the social fabric around the globe. The expanding popularity of social networking sites,[6] dating sites,[7] virtual online gaming,[8] and popular gaming consoles[9] enable unprecedented social interaction across the globe. As Iran discovered in its 2009 post-presidential elections, social networking mediums can be used to incite the population and make it difficult for governments to control their use or attribute culpability. Such mediums have vastly different technologic underpinnings, which makes them difficult to mitigate.

d. Science, technology and engineering help shape the OE and the evolution of ICT will continue to accelerate in the future. Scientific advances are poised to redefine many dimensions of society. ICT, electronics, biocomputing, and nanotechnology may profoundly affect military operations in the coming years. Developments in quantum computing and nanotechnology may lead to a fighting force enhanced by robotics and remotely guided, autonomous, and miniaturized weapons systems. Communications systems may be self-organizing and distributed. Directed energy weapons will likely be employed against high-tech U.S. systems. This means the Army must be prepared to use cyber-enabling capabilities such as spectrum management and electronic protection (EP) to guard cyber assets.

e. Operations in cyberspace can occur nearly instantaneously. Army forces can attack or be attacked with a speed not achievable in the other domains. Depending on the degree of interconnectivity, this can happen over global distances at near the speed of light. The speed in which these activities can take place poses a requirement for speed of decisionmaking heretofore never known.

f. Cyberspace has a wide range of actors with different levels of education, training, skills, motivation, and capacity. Nation-states, state-sponsored operators, nonstate actors, legitimate businesses, criminal organizations, and individuals are among these actors. The TRADOC

---

[4] Global commons is that which no one person or state owns or controls and which is central to life.
[5] ICT is the commercial equivalent of information technology used extensively in the TRADOC G-2 Operational Environment 2009-2025.
[6] MySpace and Facebook, for example.
[7] Match.com and YahooPersonals, for example.
[8] Massive multiplayer online role playing games, for example.
[9] Xbox, Playstation, and Wii, for example.

Operational Environment 2009–2025 makes it clear that operating within and in defense of this global commons will become a part of every military operation. This significant change in the OE challenges traditional understandings of military action.

g. A wide range of actors use ICT and advanced technologies as a relatively inexpensive way to gain parity with the U.S. as compared to buying tanks and aircraft or training thousands of soldiers. Therefore, ICT is a very attractive alternative for adversaries because the return on investment to create a capability is significant. Since many adversaries do not have the capability or desire to develop support structures for these capabilities, their timeline from acquisition to fielding can be significantly shorter than the U.S. acquisition systems. For cost and survival issues, the return on investment for ICT and advanced technologies makes this the only real avenue many adversaries have to maintain parity or get ahead of the U.S., hence, the adversaries' commitment to exploit this avenue.

h. Private industry research and development in large part will be the catalyst for changes in CyberOps. Since cyberspace is created, owned, maintained, and operated by public, private, and government stakeholders across the globe (figure 2-2), effective public-private partnerships will be increasingly critical to the future force. A substantial portion of what is often referred to as the "Army network" or "Army cyberspace" is actually owned and operated by commercial entities and shared by the general public. For example, mobile devices, such as cell phones and wireless personal digital assistant, are components of both the military and commercial wireless provider's networks. Leased long haul connectivity between Army installations including fiber optic cabling, routing, and switching through the physical infrastructure is shared between military and civilian networks. There are leased and managed services in which military information resides on commercial devices. A detailed public-private partnership that includes roles, responsibilities, and authorities needs to be developed because these commercial segments are a critical part of the Army's network. This creates several significant challenges for the Army - both operationally and in the development of DOTMLPF solutions.

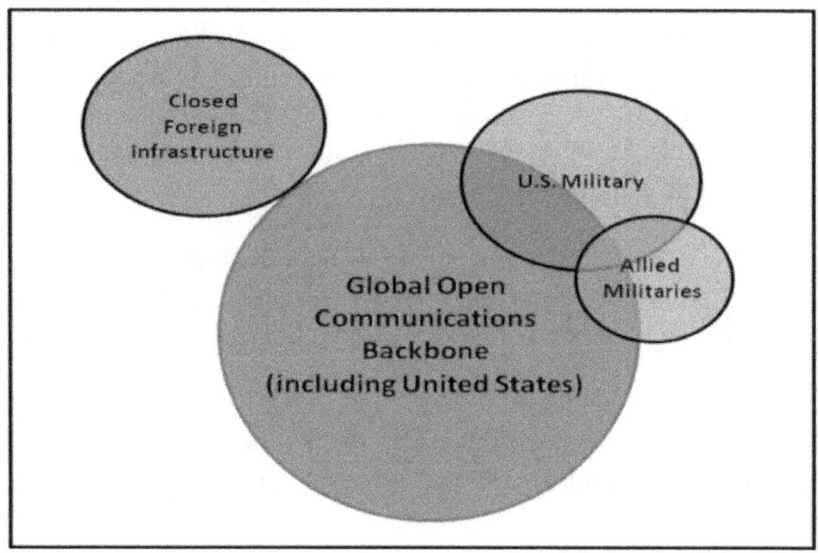

Figure 2-2. Cyberspace connectivity

11

i. The Army depends upon the Nation's critical infrastructure and key resources for many of its activities, including force deployment, training, transportation, and normal operations. Physical protection of these is no longer sufficient as most critical infrastructure is controlled by networked and interdependent SCADA or distributed control systems (DCS). The Department of Homeland Security (DHS) chart at figure 2-3 highlights the various infrastructures that must be protected.

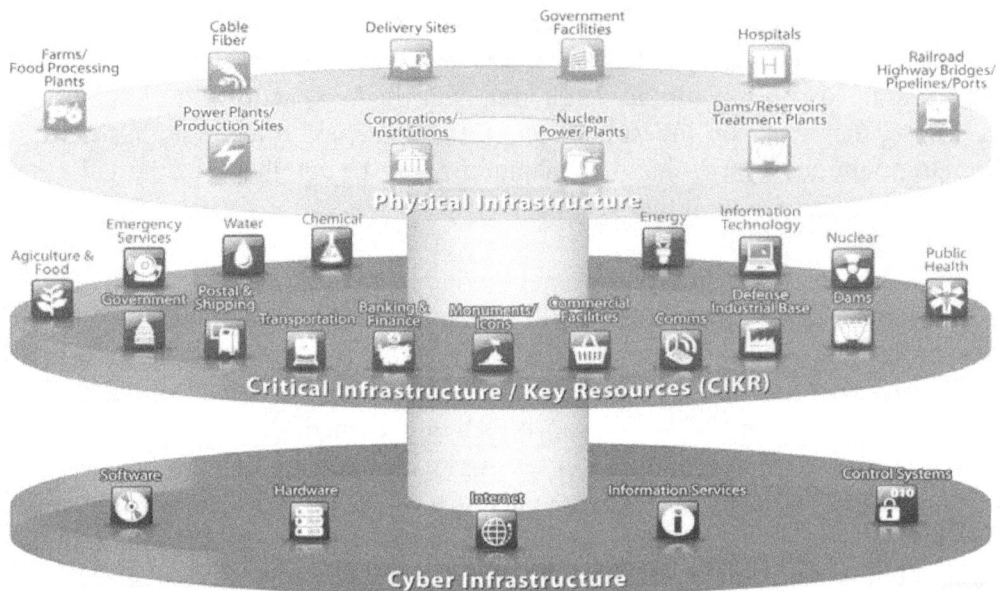

Figure **2-3. Infrastructure relationships in cyberspace**[10]

j. Since private industry is the primary catalyst for technologic advancements, the military may become increasingly reliant on commercial off-the-shelf (COTS) technology. This reliance may present three primary vulnerabilities:

(1) Foreign ownership, control, and influence of vendors. Many of the COTS technologies (hardware and software) the Army purchases are developed, manufactured, or have components manufactured by foreign countries. These manufacturers, vendors, service providers, and developers can be influenced by adversaries to provide altered products that have built in vulnerabilities, such as modified chips.

(2) Supply chain. The global supply chain has vulnerabilities that can potentially lead to the interception and alteration of products. These vulnerabilities are present throughout the product life cycle, from the inception of the design concept, to product delivery, and to product updates and support.

(3) COTS and government off-the-shelf (GOTS) balance. The vast majority of the Army's CyberOps components and capabilities are from COTS and to a much smaller degree, GOTS technologies.

---

[10] Source: DHS, "Securing the Nation's Critical Cyber Infrastructure."

2-3. The cyber threat

a. The Army is facing multiple, simultaneous, continuous, hybrid threats that employ regular and irregular forces and use an ever-changing variety of conventional and unconventional tactics to achieve their aims. This diverse and wide variety of threats has disparate agenda, alliances, and range of capabilities. These actors include traditional nation-states, noncombatants, transnational corporations, criminal organizations, terrorists, hacker unions, mischievous hackers, and the unwitting individual who intends no malice. The Army must be able to deal with this wide array of threats.

b. The cyber threat can be characterized in many ways: by sponsorship, training, education, skills, motivation, or tools. Two examples include advanced cyber threats and hackers. Advanced cyber threats are generally supported by nation-states and have advanced education, training, skills, and tools that allow the threat to remain undetected for extended periods of time on improperly defended networks. Hackers have a broad range of skills, motives, and capabilities and must be assessed independently. The level of the cyber threat is the combination of the actor's ability (skills and resources), opportunity (access to target), intent (attack, surveillance, exploit), and motive (national policy, war, profit, fame, personal reasons, and others).

c. With access to the vastly available commercial tools and distribution methods, adversaries have proven adept at rapidly adjusting to evolving commercial technology; they have cleverly adopted new methods to reach their intended actors and publics. With modest resources, they can purchase dual use civil technologies making the marketplace their combat developer. When compared to current DOD requirements and acquisition processes, which take multiple years to adopt new technologies across the force, adversaries are able to leap ahead and achieve a significant advantage.

d. Cyberspace provides adversaries an effective and inexpensive means for recruitment, propaganda, training, and command and control. Increasingly, nations and nonstate actors will use cyberspace effectively, often in combination with lethal attacks and an information campaign, to advance their aims. In Iraq, ICT became an essential ideological and operational tool for the insurgency. Many groups carried out sophisticated information strategies with video cameras, laptops, cell phones, and other wireless technologies. This trend is likely to evolve with future threats operating in and through cyberspace to delay or disrupt U.S. access to the theater or area of operations by striking key infrastructures, such as aerial and sea ports of debarkation and embarkation, lines of communication, staging bases, domestic and SCADA systems, and sea and air transports.

2-4. Summary

a. Operations conducted in and through cyberspace will have significant impact on Army missions worldwide. Small, nimble adversaries have proven committed to adapting cyber-electromagnetic technologies very quickly due to the huge return on their investment and their ability to gain operational parity or advantage, even if fleeting. U.S. telecommunication and computer networks are targeted, engaged and/or attacked, and defended continuously each day.

These activities will increase in the future due to the accelerating rate of technologic advances, the low entry cost for commercial technology, its resulting proliferation, and the ability adversaries, even unsophisticated, to stay apace of the increasing rate of technologic change.[11]

b. Notions of "dominating" cyberspace are simplistic and unrealistic. A realistic and meaningful goal is to achieve and maintain freedom of action in and through cyberspace while being able to affect that of the adversaries. Subsequent chapters will begin to determine the cyber capabilities and capacity required to prevail in the cyber-electromagnetic contest with current and future adversaries.

Chapter 3
Framing the Problem

3-1. What has changed in the operational environment as pertains to cyberspace?
As detailed in chapter 2, the future operational environment continues to change in accelerated fashion due to ICT advancements and the application of these technologies. Global proliferation of these technologies has significantly changed social networks and the social dynamic. Adaptive and flexible adversaries have proven adept at leveraging these technologies and adapting them to this environment. Cyberspace has emerged as a realm of continuous engagements and conflict where there is no peacetime and no clear cut winners or end states.

3-2. What has changed in operational requirements as pertains to cyberspace?
The Army has placed significant emphasis and reliance on network-enabled systems and services to provide a communications infrastructure that connects Soldiers and platforms to global information sources, which are increasingly susceptible to attack, degradation, and/or destruction. CyberOps are conducted continuously to combat the array of adversaries attacking friendly systems and to ensure the access to and operation of Army and other specified cyberspace networks. Current requirements do not meet and/or address the increased speed, massive proliferation of information, and access to relevant information in cyberspace. Joint and Army cyberspace requirements have increased due to the expanding mission area, diverse threats, and with the establishment of the U.S. Cyber Command (USCYBERCOM) and U.S. Army Forces Cyber Command (ARFORCYBER).

3-3. What is working, what is not working?

a. What is working? The Army has several organizations that are at the forefront in training Army Soldiers and civilians to provide commanders with personnel who possess the right knowledge, skills, and abilities to perform cyber activities, CNO, and NetOps in particular.

b. What is not working? The Army currently does not have a holistic vision, concept, or doctrine to guide its capability development efforts in response to the changes in the OE and operational requirements for CyberOps. There has been no comprehensive analysis to determine CyberOps requirements and to guide its development and management of CyberOps capabilities

---

[11] "*Cyberpower and National Security*", Chapter 1, written by Franklin D. Kramer (National Defense University).

across DOTMLPF. The Army has insufficient research, development, test and evaluation (RDT&E) funding to responsively support current and future CyberOps.

3-4. The problem
The Army cannot adequately identify, attack, exploit, and defeat the expanding cyber-electromagnetic threats or mitigate the increasing vulnerability of its own networks. The Army is not poised to prevail in the cyber-electromagnetic contest.

Chapter 4
Framing the Solution

4-1. The context: The three dimensions of FSO

a. The Army has considered the role of information, cyberspace, and the electromagnetic spectrum since the publication of FM 3-0. The staffing for both the draft FM 3-13 and the "U.S. Army Cyber-Electronics Concept of Operations" revealed significant and different opinions about how the Army should be organized for cyberspace, EW, and IO. Consequently, the CG TRADOC directed the CAC to lead a working group to recommend the conceptual framework for these mission areas and the needed force modernization proponency structure. The CAC-led working group conducted two Councils of Colonels and a General Officer Review Board to accomplish these two tasks.

b. The group leveraged the design process to develop an environmental frame, a problem frame, and a solution frame. The frames derived three dimensions of FSO and the General Office Review Board validated them as logical and good enough to move forward while the group continued to learn and reframe the problem. On 16 October 2009, CG TRADOC provided his recommendations to the Army Vice Chief of Staff in a memorandum titled, "Posturing the Army for Cyber, EW, and IO as Dimensions of Full Spectrum Operations." CG recommendations included the following.

(1) "We determined that our current 'vocabulary' (Cyber-EW-IO) is adequate today but will become increasingly inadequate to describe the challenges we face in the operational environment. We concluded that there are three dimensions to be addressed, that these dimensions exist across the FSO, and that these dimensions each require force design and doctrinal solutions. Therefore, although we currently describe the functions related to network and spectrum operations as Cyber-EW-IO, we believe that we should adapt and describe them in the future as follows:

(a) First dimension. The first dimension is the psychological **contest of wills** against implacable foes, warring factions, criminal groups, and potential adversaries. This dimension involves influencing desperate and creative people "to do what they really don't want to do" and requires an acute understanding of human behavior.

(b) Second dimension. The second dimension is **strategic engagement** and involves keeping friends at home, gaining allies abroad, and generating support or empathy for the mission in the area of operations. This dimension includes the general public, key actors, and third party validators who are the ultimate arbiters of success or failure of military operations in the current operational environment. Gaining and maintaining their support or empathy for the mission is an imperative of 21st century operations.

(c) Third dimension. The third dimension is the **cyber-electromagnetic**[12] **contest**. Trends in wired, wireless, and optical technologies are setting conditions for the convergence of computer and telecommunication networks. A significant advantage will go to the side that is able to gain, protect, and exploit advantages in the highly contested cyberspace and electromagnetic spectrums."

(2) The future force must meet the demands of these three interconnected dimensions of FSO in an operational environment characterized by complexity, rapid change, and hybrid threats, this pamphlet is about prevailing in the third dimension.

c. Winning the cyber-electromagnetic contest often determines to a large degree the capability of military forces to perform missions. This is not some metaphorical cyberspace. This is a dimension shaped and bounded by: modern computer-mediated communications networks of all types; transmission of data within networks by electromagnetic waves, fiber optic cables and copper wire; digital information storage and processing; computerized automation; a large variety of sophisticated electronic sensors; space-based communications, broadcast, mapping, and global positioning services; various electronically activated remote control systems; and other rapidly evolving network services and applications.

d. Rapidly evolving information technologies are increasing the speed, capacity, agility, efficiency, and usefulness of modern networks. The proliferation of this technology is changing the way humans interact with each other and their environment, to include military operations. The U.S. Army is heavily reliant on information technology and information systems to communicate, control forces, coordinate fires, gather and distribute intelligence, and conduct surveillance, reconnaissance, and other military activities. U.S. adversaries, warring factions, and criminal cartels have access to and use many of the same technologies in innovative ways that are unique to every case.

e. How these cyber-electromagnetic technologies are integrated and employed in specific circumstances will greatly affect modern military operations. While it is important to be at the leading edge in these technologies, it is just as important to take a comprehensive approach to all aspects of this dimension of operations and to be the cleverest to adapt and combine them to unique operating conditions.

4-2. Central idea

a. Prevailing in the cyber-electromagnetic contest means making progress at the same time along three lines of effort: gaining advantage, protecting that advantage, and placing adversaries at a disadvantage.

---

[12] The use of the term cyber-electromagnetic is not meant to equate the terms cyberspace and electromagnetic spectrum, but rather to highlight there is significant overlap between the two and future technological development is likely to increase this convergence.

b. While there is great advantage in harnessing cutting edge ICT ahead of adversaries, implementation must consider and mitigate emerging systemic vulnerabilities and dependencies. Perhaps even more critical is the ability to disarm, disrupt, and defeat the same capabilities in the hands of adversaries. This requires Army forces to integrate these lines of effort from the start, making them elements of the same dimension of modern operations. Integration leads to synergy, rapid progress, and high relative levels of performance. Failing to integrate leads, at best, to uneven progress and disjointed applications, or at worst, catastrophic operational failures.

c. The art of winning in the cyber-electromagnetic dimension requires very specific expertise in information theory, computer science, and related sciences (electro-physics, radio-electronic wave propagation theory, cyber-electronics, complex cyber network behaviors, and others) and of how this theoretical knowledge relates to military tactics, operations, and strategy.[13] Creating this marriage of abstract science and modern military practice is fundamental to creating CyberOps SA and thus contributing to the commander's end state. Another is to transform the fragmented approach to this dimension into one that is systemically holistic. Gaining advantage and denying advantage through modern, high technology, automation-enhanced networks depend on the same scientific knowledge base, and are symmetrically related aspects of the same contest.

d. While it is possible military outcomes can be determined by cyber operations alone, CyberOps are not generally an end to themselves but rather an integral part of FSO. It is focused on winning the cyber-electromagnetic contest through three concurrent lines of effort: gaining advantage, protecting that advantage, and placing adversaries at a disadvantage. Commanders conduct CyberOps to retain freedom of action in cyberspace and in the EMS, while at a time and place of their choosing, denying freedom of action to adversaries, thereby enabling other operational activities. These lines of effort to prevail in the cyber-electromagnetic contest nest with and contribute to the joint force's construct of cyberspace superiority. CyberOps leverages cyberspace and the EMS throughout all the domains.

4-3. The framework

a. Current doctrinal terms do not adequately address the broad range of tasks associated with the DOD definitions of cyberspace and CyberOps. For example, cyberspace includes computer and telecommunication networks as well as embedded processors and controllers in equipment, systems, and infrastructure; and CyberOps encompasses more than just CNO and NetOps. Consequently, the framework developed for this pamphlet establishes four components for CyberOps: CyberSA, CyNetOps, CyberWar, and CyberSpt, with CyberWar and CyNetOps being the primary operational components. This framework is illustrated in figure 4-1 and further elaborated upon in this chapter.

---

[13] "Just as it is necessary to understand human psychology and human social behavior to succeed in the art of unifying physical and psychological impact, and that or keeping friends and winning allies, knowledge in these fields is crucial to this art. The first term, electrophysics, is the root science that defines this field. Cyber-electronics is a term I prefer over Cyberspace to cover the science that bounds and defines modern communications, including the Internet. Moreover, the character of modern operations is so shaped by these sciences, and the enabling capabilities that stem from them, that to not consider these a "dimension" would be limiting." *Introduction to Winning in the Cyber Electromagnetic Dimension of "Full Spectrum Operations,"* Brigadier General Huba Wass de Czege, U.S. Army, Retired.

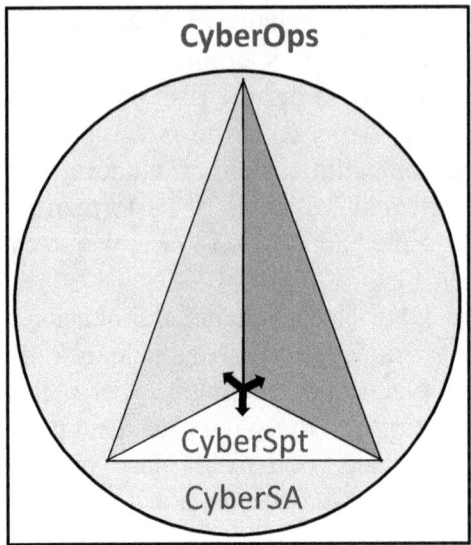

Figure 4-1. The four components of CyberOps

b. CyberSA is the immediate knowledge of friendly, adversary and other relevant information regarding activities in and through cyberspace and the EMS. It is gained from a combination of intelligence and operational activity in cyberspace, the EMS, and in the other domains, both unilaterally and through collaboration with unified action and public-private partners. Discrimination between natural and manmade threats is a critical piece of this analysis. CyberSA enables informed decisionmaking at all levels via flexibly tailored products and processes that can range from broadly disseminated awareness bulletins targeted to general users to the other extreme of specific and narrowly focused issues distributed as extremely sensitive and classified in nature. CyberSA enables informed decisionmaking at all levels. It is relevant at the strategic, operational, and tactical echelons for overall SA; and it is useful to Soldiers who interact most with the populace, which uses and increasingly relies on cyberspace. As depicted in figures 4-1 and 4-2, CyberSA enables and derives from CyNetOps, CyberWar, and CyberSpt. CyberSA includes the following.

(1) An understanding of friendly, adversary, and other relevant activity in and through cyberspace.

(2) Assessment of friendly cyber capabilities.

(3) Assessment of adversary cyber capabilities and intentions.

(4) Assessment of both friendly and adversary cyber vulnerabilities.

(5) An understanding of information flowing over networks to include its purpose and criticality.

(6) An understanding of the effects and mission impact resulting from friendly and adversary cyberspace degradations.

(7) Availability of cyber capabilities necessary for the effective planning, and execution of CyberOps.

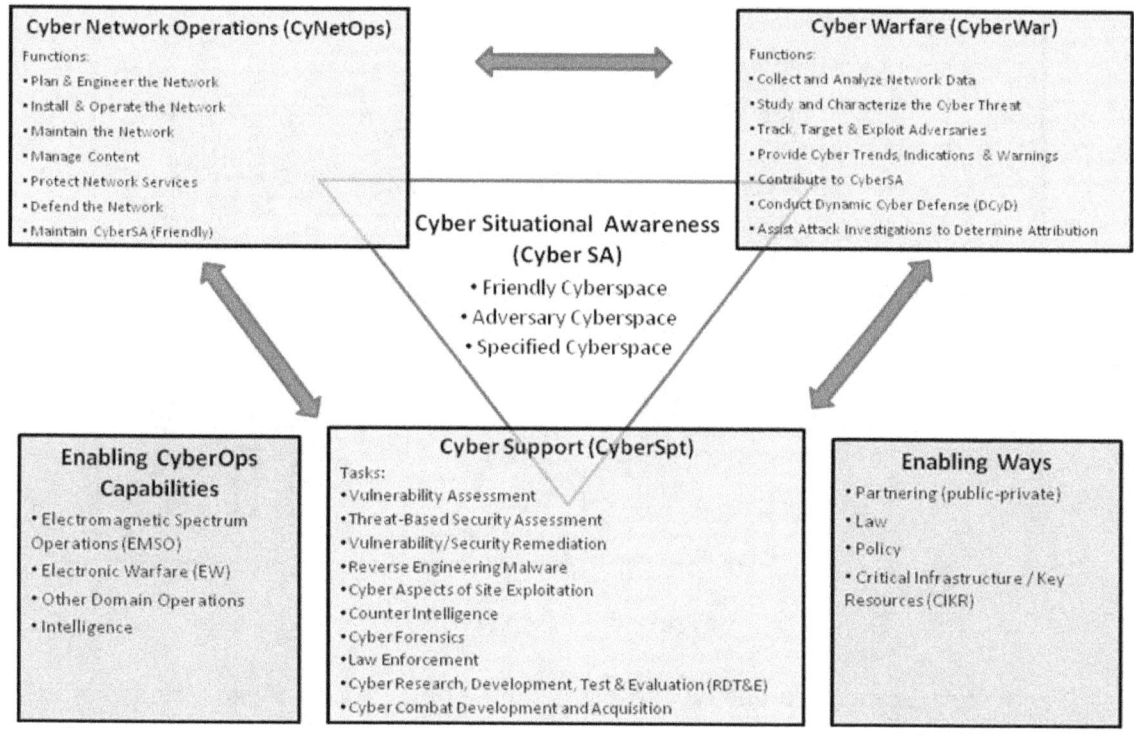

Figure 4-2. CyNetOps

c. CyNetOps is the component of CyberOps that establishes, operates, manages, protects, defends, and commands and controls the LandWarNet[14], critical infrastructure and key resources (CIKR), and other specified cyberspace. CyNetOps consists of three core elements: Cyber enterprise management (CyEM), cyber content management (CyCM), and cyber defense (CyD), including information assurance, computer network defense (to include response actions), and critical infrastructure protection. CyNetOps uses CyEM, CyCM, and CyD in a mutually supporting and supported relationship with CyberWar and CyberSpt (see figure 4-3).

---

[14] LandWarNet is the Army's contribution to the GIG that consists of all globally interconnected, end-to-end set of U.S. Army information capabilities, associated processes, and personnel for collecting, processing, storing, disseminating, and managing information on demand supporting warfighters, policy makers, and support personnel. It includes all U.S. Army (owned and leased) and leveraged DOD and joint communications and computing systems and services, software (including applications), data security services, and other associated services. LandWarNet exists to enable the war fight through battle command. (TRADOC Pam 525-5-600).

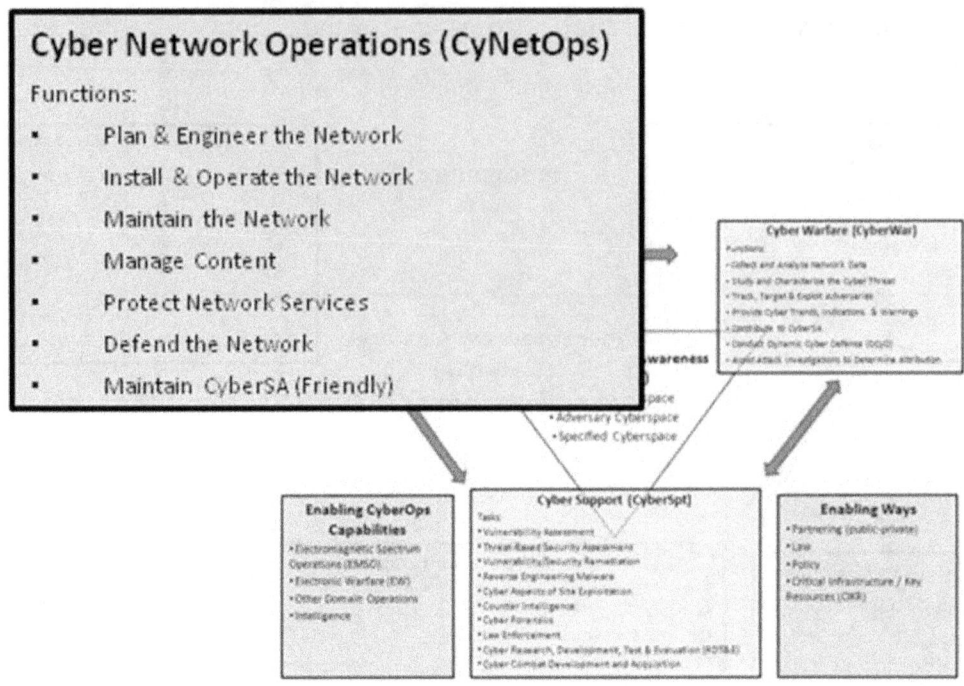

Figure 4-3. CyNetOps

(1) CyEM is the technology, processes, and policy necessary to operate effectively computers and networks.

(2) CyCM is the technology, processes, and policy necessary to provide awareness of relevant, accurate information; automated access to newly discovered or recurring information; and timely, efficient, and assured delivery of information in a usable format.

(3) CyD actions combine information assurance, computer network defense (to include response actions), and critical infrastructure protection with enabling capabilities (such as EP, critical infrastructure support, and others) to prevent, detect, and ultimately respond to an adversaries ability to deny or manipulate information and/or infrastructure. CyD is integrated with the dynamic defensive aspects of CyberWar to provide defense in depth.

(4) The rapidly changing nature of cyberspace mandates that operational and tactical units possess organic, or have access to, the capabilities and expertise to protect these vital networks; enable real time attack prevention and detection; make possible attack response through event identification and actions such as deception, blocking and/or denying; and allow the coordination of appropriate counterattacks.

(5) The availability of information and intelligence via the LandWarNet and other aspects of cyberspace are critical to all operations and overall mission success. The defenses and network redundancies must be sufficiently robust to provide security and continued availability in spite of the adversary's attempts to exploit or attack critical systems and networks. Potential adversaries possess significant CyberOps capabilities and Army forces will likely have to fight through a threat event while operating in a degraded environment, especially at the operational

and tactical levels. In response to this threat, the Army must train to operate with degraded systems.

d. CyberWar is the component of CyberOps that extends cyber power beyond the defensive boundaries of the GIG to detect, deter, deny, and defeat adversaries. CyberWar capabilities target computer and telecommunication networks and embedded processors and controllers in equipment, systems and infrastructure. CyberWar uses cyber exploitation (CyE), cyber attack (CyA), and dynamic cyber defense (DCyD) in a mutually supporting and supported relationship with CyNetOps and CyberSpt (see figure 4-4).

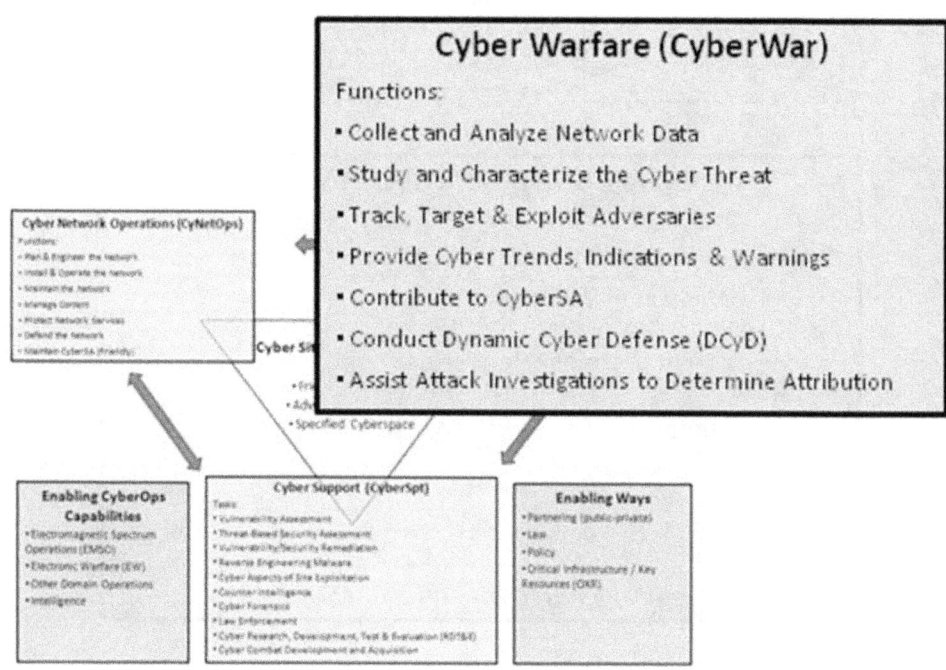

Figure 4-4. CyberWar

(1) CyA actions combine computer network attack (CNA) with other enabling capabilities (such as, electronic attack (EA), physical attack, and others) to deny or manipulate information and/or infrastructure.

(2) CyE actions combine computer network exploitation (CNE) with enabling capabilities (such as, electronic warfare support (ES), signal intelligence (SIGINT), and others) for intelligence collection and other efforts.

(3) DCyD actions combine policy, intelligence, sensors, and highly automated processes to identify and analyze malicious activity, simultaneously tip and cue and execute preapproved response actions to defeat attacks before they can do harm. DCyD uses the Army defensive principles of security, defense in depth, and maximum use of offensive action to engage cyber threats. These actions include surveillance and reconnaissance to provide early warnings of pending enemy actions. DCyD is integrated with the defensive aspects of CyNetOps to provide defense in depth.

e. CyberSpt is a diverse collection of supporting activities which are generated and employed to specifically enable both CyNetOps and CyberWar (see figure 4-5). These activities are called-out in this unifying category due to their unique and expensive nature as high-skilled, low-density, time-sensitive/intensive activities requiring specialized training, processes, and policy. Additionally, several of these activities also require specialized coordination, synchronization, and integration to address legal and operational considerations. It is because of these considerations and their overall importance that these activities are addressed as a CyberOps core component.

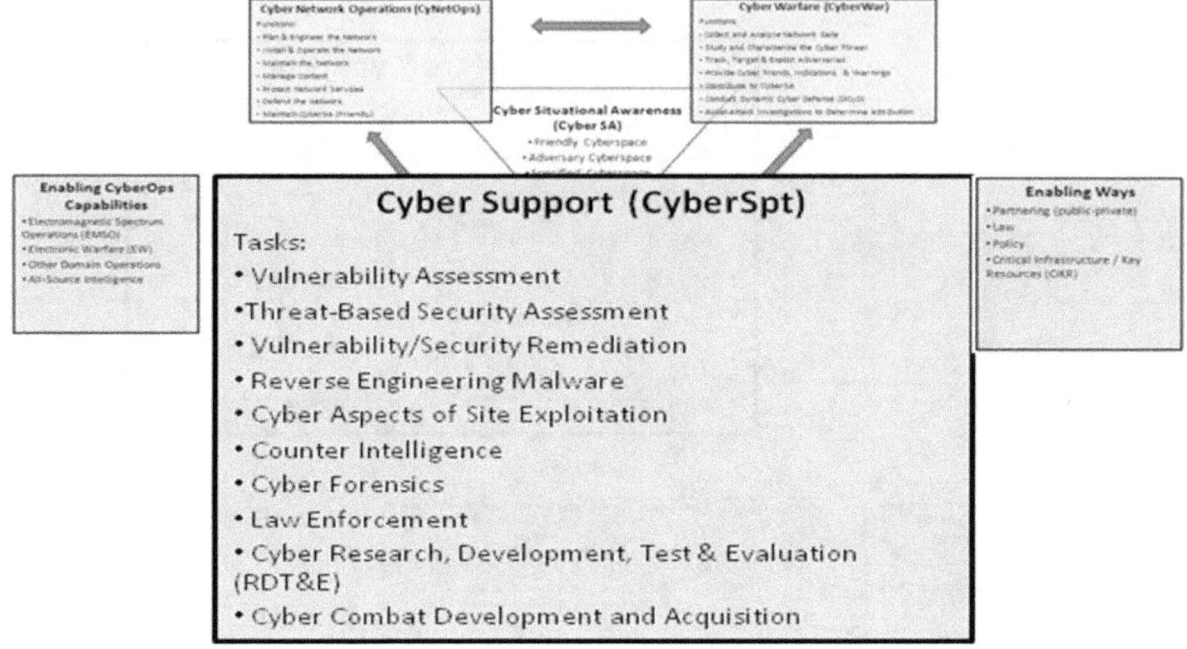

Figure 4-5. Cyber support

(1) CyberSpt is different from CyberWar and CyNetOps, as these activities are carried out by multiple stakeholders and do not require a separate CyberSpt proponent or lead. At quick glance, ownership may seem clouded as intelligence organizations, law enforcement, CyberWar, and CyNetOps perform many of these activities. CyberSpt entails varying intents, conditions, authorities, and levels of effort and are not deemed redundant.

(2) Examples of CyberSpt activities include vulnerability assessment, threat-based security assessment, and remediation; reverse engineering malware; cyber aspects of site exploitation; separate counterintelligence and law enforcement-based cyber forensics; cyber RDT&E; combat development; and acquisition. These are low density, high demand capabilities that must be expanded to support emerging requirements.

(3) Particularly noteworthy is a responsive RDT&E and science and technology strategy carried out by organizations that reside in the Army and integrated with other services, DOD, and other governmental agencies, in industry and academia. The goal is to identify and evaluate promising technologies before they emerge in the marketplace. The Army must invest in and

leverage these organizations to stay apace of commercial technologic advancements and to prevent the introduction of game-changing technologies by adversaries. Due to the cost of cyber developments, it is likely much of this development will be coordinated and funded through the USCYBERCOM. It will be increasingly important the Army assures the development strategy meets its specific requirements.

f. Enabling CyberOps capabilities.

(1) Army operations rely on systems that use the EMS to conduct FSO. The convergence of computer and telecommunications networks and the proliferation of advanced technologies make it imperative that CyberOps and EW are deconflicted, fully coordinated, and synchronized with all other aspects of the operation to achieve the commander's intent and objectives. Enablers enhance the effectiveness and integration of military capabilities and their subsequent effects.

(2) Due to the competitive and congested environment, access to cyberspace and the EMS cannot be assumed. Host nations are unlikely to have the sophistication or capacity for radio frequency (RF) spectrum management required for coalition operations. Army forces should be prepared to supplement host nation capabilities. EMSO provide these spectrum management, frequency assignment, host nation coordination, and policy implementation capabilities that are essential to gain the required access to enable CyberOps. Figure 4-6 shows that EMSO is an enabling CyberOps capability along with electronic warfare, operations in other domains, and intelligence.

(3) The reliance of CyberOps on RF and optical portions of the EMS make EW another enabler. EW activities represent a distinct enduring capability to provide Army commanders an advantage within the EMS. These commanders must have a similar CyberOps capability to attack, protect, and exploit advanced technologies. The combination of CyberOps and EW provides the commander with alternative authorities and multiple techniques to achieve desired effects under varying conditions. A CyA for example, could provide additional opportunity to disrupt adversary's communications beyond those offered by EA but using the same RF spectrum. CyD may provide additional protection to data transmission devices beyond the shielding provided by EP capabilities. And, CyE can enhance target development and intelligence information beyond what is currently offered by ES capability. A properly designed platform may be able to deliver both capabilities in one system.

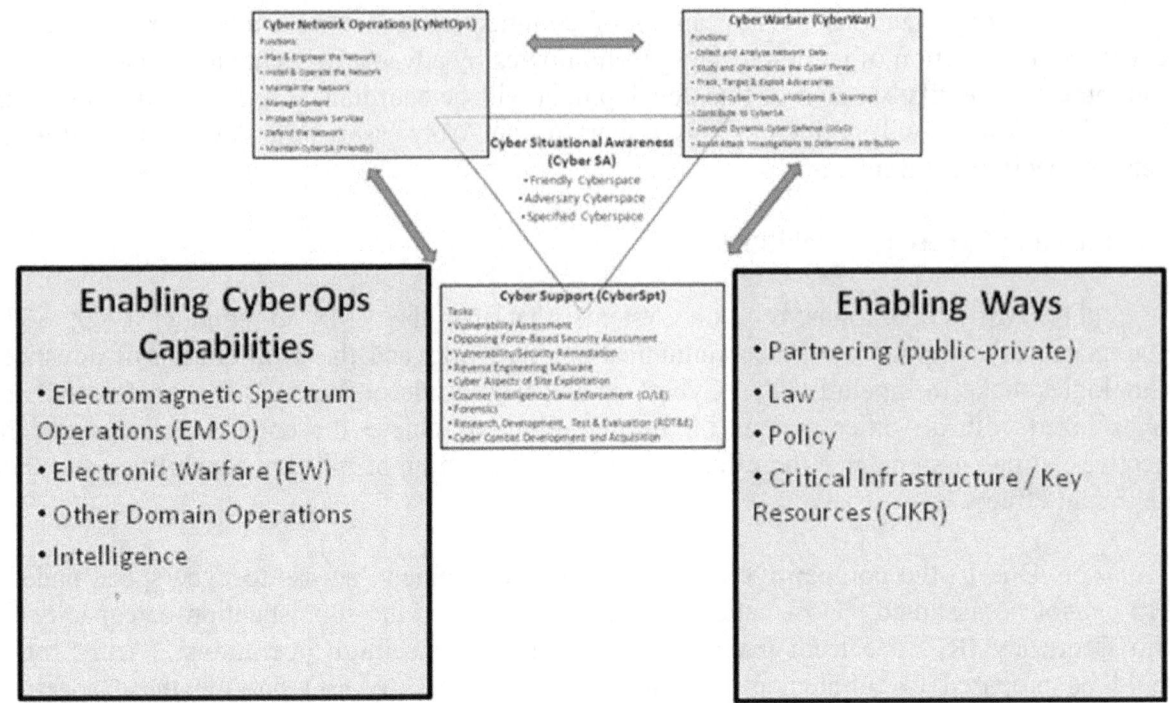

Figure 4-6. Enabling **CyberOps capabilities and enabling ways**

(4) Capabilities normally employed in the air, land, maritime, and space domains can also enable CyberOps and vice versa. Space capabilities are employed to extend the Army's LandWarNet and to deliver attack and exploitation payloads to adversary' systems and networks. Aerial and high altitude platforms provide additional means to accomplish many of the same tasks performed by space capabilities in areas where more responsive and persistent coverage is needed. Similarly, physical attack and other capabilities in the land domain can enable CyberOps.

(5) Intelligence. SIGINT and other intelligence activities enable, and are enabled by, all aspects of CyberOps in both support and operational contexts. Increasingly technologically advanced adversaries require that U.S. intelligence and cyber personnel work closely together, using an all source approach, to support the commander's mission and to build CyberSA.

g. Enabling ways. Enabling ways enhance the effectiveness and integration of CyberOps capabilities. Partnering in unified action and CIKR include continuous actions taken to shape the operational environment and set the conditions for successful operations. Law and policy authorize or place limitations on what can be done operationally and feed the development of rules of engagement (ROE). Each is described below:

(1) Partnering. CyberOps require interdependent capabilities and unified action and there are multiple partnering arrangements that must be made to ensure effective operations. Public-private partnerships are essential because the vast majority of cyberspace is owned and controlled by public and private stakeholders. The Army must have established enduring partnerships with Federally-funded research and development centers, industry, other services,

academia, commercial infrastructure providers, defense contractors, and other global and regional allies and partners who share similar dependence on cyberspace. Partnering with interagency organizations is an important aspect of unified action in this mission area. In addition, partnering with Federal, state, and local law enforcement, counterintelligence, and criminal investigative organizations will facilitate the rapid coordination needed for effective CyberOps, including the pursuit, investigation, and prosecution of criminals.

(2) CIKR. The Army depends upon the Nation's CIKR for its day-to-day operations, transportation, power, and information and communications technology infrastructures. CIKR is also important for the deployment and training of Army forces. Physical protection of CIKR is no longer sufficient as most critical infrastructure is controlled by networked, interdependent SCADA or DCS. The Army depends upon established partnerships and must be prepared to conduct both cyber and physical CIKR protection as part of FSO.

(3) Law and policy. CyberOps and enabling capabilities are governed through a complex set of classified and unclassified legislation, policy, and procedures, and ROE resulting in overlapping authorities among DOD, the Federal Bureau of Investigation, DHS, Department of State, Department of the Treasury, and other government agencies. Law, policy, and ROE are distinct entities that authorize actions and also place restrictions and limitations on what can be done operationally. These are critical for effective operations and for the proper oversight of these operations. Presidential National Security Presidential Directives, Homeland Security Presidential Directives, and the 2008 Unified Command Plan provide policy guidance on CyberOps at the National level. Relevant cyber U.S. Code includes Title 6 for Domestic Security; Title 10 for the Armed Forces; Title 18 for Crimes and Criminal Procedure; Title 32 for the National Guard; Title 40 for Public Buildings, Property, and Works; and Title 50 for War and National Defense. Additionally, there are numerous DOD, Joint Staff, and service policies on CyberOps; and, the authority to conduct CyberOps against an adversary not located within the U.S. may be impacted by the United Nations and North Atlantic Treaty Organization charters and other treaties. Laws, policies, and authorities for CyberOps must be understood clearly and relationships established accordingly to facilitate rapid execution of operations.

4-4. Risks

a. As the vignettes in appendix B show, the U.S. Army may be required to augment host nation and civil support agencies with CyberOps expertise and capabilities. The vignettes posit the joint force will provide this augmentation to Army forces since it will exceed the Army's capacity. However, at present such is far from being a reality. Failure to build this capacity in the joint force will place both mission and lives at risk.

b. The second category of risk is technologic in nature. The assumption for successful integration of CyberOps into FSO is that the DOD and Army will pursue in earnest a competitive advantage in CyberOps capabilities. Failure to adapt research, development, testing, and acquisition processes to stay apace with technologic advancements will make it difficult, if not impossible, to gain advantage, protect that advantage, and place adversaries at a disadvantage.

4-5. Summary

a. Trends in wired, wireless, and optical technologies are setting conditions for the convergence of computer and telecommunication networks. Winning the cyber-electromagnetic spectrum dimension of FSO determines to a large degree the capability of military forces to perform missions of all kinds. A significant advantage will go to the side that is able to gain, protect, and exploit advantages in the highly contested cyberspace and EMS. Prevailing in the cyber-electromagnetic contest means making progress at the same time along three lines of effort: gaining advantage, protecting that advantage, and placing adversaries at a disadvantage, from peacetime engagements to global war.

b. CyberOps encompass those actions aimed at gaining advantage, protecting that advantage, and placing adversaries at a disadvantage in cyberspace and in the EMS, just as commanders do in and across the air, land, maritime, and space domains. Commanders seek to retain freedom of action in the cyberspace and EMS, while denying the same to their adversaries, thereby enabling other operational activities in and through cyberspace as well as in the other four domains. CyberOps are not an end to themselves but rather an integral part of FSO that focus on winning the cyber-electromagnetic contest by gaining advantage, protecting that advantage, and placing adversaries at a disadvantage. CyberOps use cyberspace and the EMS and take place in the air, land, maritime, and space domains as well as in and through cyberspace.

c. CyberOps uses four components, along with enabling capabilities and special considerations to achieve the commander's intent. The four components of CyberOps, CyberSA; CyNetOps; CyberWar; and CyberSpt, are interdependent and must be integrated into the commander's overall operation.

d. The Army's ability to leverage cyberspace and CyberOps capabilities will be increasingly critical to its operational success. CyberOps capabilities must be fully integrated in right combination with all other capabilities at the commander's disposal to gain advantage, protect that advantage, and place adversaries at a disadvantage. To do this, the Army must possess the required cyber capabilities across DOTMLPF domains and provide them to USCYBERCOM, combatant commanders, and Army operating and generating forces. This pamphlet's appendices build on the framework in the base document to broaden the operational understanding of the framework and to provide a start point for the CBA. This pamphlet provides a vision for the Army's way forward in the development of CyberOps capabilities to prevail in the cyber-electromagnetic contest with current and future adversaries.

Appendix A
References

Section I
Required References
ARs, DA Pamphlets, field manuals (FM), and DA forms are available at Army Publishing Directorate Home Page http://www.usapa.army.mil. TRADOC publications and forms are available at TRADOC Publications at http://www.tradoc.army.mil.

No Entry

Section II
Related References

Army Posture Statement.

Army Strategic Planning Guidance, FY2006-2023.

Army Transformation Roadmap

Capstone Concept for Joint Operations.

CJCSM 3500.04D
Universal Joint Task List.

Command and Control Joint Integrating Concept.

Department of Defense Information Enterprise Architecture.

Department of the Army Training Manual 5-601
Supervisory Control and Data Acquisition Systems for Command, Control, Communications, Computer, Intelligence, Surveillance, and Reconnaissance Facilities.

Field Manual 3-0
Operations.

Field Manual 3-13
Information Operations.

Field Manual 3-36
Electronic Warfare Operations.

Field Manual 6-02.70
Electromagnetic Spectrum Operations.

Field Manual 6-02.71
Network Operations.

FM 7-0
Training for Full Spectrum Operations.

FM 7-15
The Army Universal Task List.

Homeland Defense and Civil Support Joint Operating Concept.

Joint Concept of Operations for Global Information Grid NetOps.

Joint Operating Environment 2008.

Joint Publication 3-0
Joint Operations.

Joint Publication 3-13
Information Operations.

Major Combat Operations Joint Operating Concept.

Military Support to Stabilization, Security, Transition, and Reconstruction Joint Operating Concept.

National Defense Strategy of the United States.

National Strategy to Secure Cyberspace

National Counterintelligence Strategy of the United States.

National Intelligence Estimate "The Global Cyber Threat to the U.S. Information Infrastructure (U)."

The Army in Joint Operations, The Army's Future Force Capstone Concept 2016-2028.

TRADOC G-2 Operational Environment 2009-2025.

TRADOC Memorandum, "Posturing the Army for Cyber, EW, and IO as Dimensions of Full Spectrum Operations."

TRADOC Pamphlet 525-3-0
The Army Capstone Concept: Operational Adaptability—Operating Under Conditions of Uncertainty and Complexity in an Era of Persistent Conflict.

TRADOC Pamphlet 525-5-600
The United States Army's Concept of Operations LandWarNet 2015.

TRADOC Pamphlet 525-7-6
United States Army Concept Capability Plan for Army Electronic Warfare Operations for the Future Modular Force 2015-2024.

TRADOC Pamphlet 525-7-16
United States Army Concept Capability Plan for Army Electronic Electromagnetic Spectrum Operations for the Future Modular Force 2015-2024.

TRADOC Pamphlet 525-7-17
The U.S. Army Concept Capability Plan for Network Transport and Services for the Future Modular Force 2015-2024.

Section III
Prescribed Forms

No entry.

Section IV
Referenced Forms

DA Form 1045
Army Ideas for Excellence Program Proposal.

DA Form 2028
Recommended Changes to Publications and Blank Forms.

Appendix B
Interdependent Nature of CyberOps

B-1. Introduction

The purpose of this section is to introduce enough of the evolving cyber operational structure to set the stage for appendix C and describe what capabilities are needed by Army echelons to conduct CyberOps. Since this operational structure is evolving, insomuch as possible, current organizational names will not be used. This section provides an overview and the context for the vignettes in appendix C. It is not intended to be a comprehensive analysis or list of organizational requirements. The Cyberspace CBA will conduct additional analysis and refine organizational requirements.

B-2. Operational overview 1 (OV-1)

a. Figure B-1 depicts the CyberOps operational overview (OV-1), a broad concept of how the future force will conduct CyberOps as part of FSO. Implicit in the illustration are the following conditions:

(1) These operations are conducted in friendly, adversarial, and other specified cyberspace with unified action partners.

(2) Public-private partnerships are paramount to the success of CyberOps operations.

(3) There are no rear areas, and CyberOps apply equally to the generating force.

(4) While some aspects of CyberOps require physical proximity, they can be conducted globally from nearly any location.

(5) CyberOps require a capability mix of organic unit capabilities and reach back to joint, Army, and interagency support organizations.

(6) The nature of the OE makes cyberspace, EMS, and the other four domains (land, air, maritime, and space) inexorably interdependent.

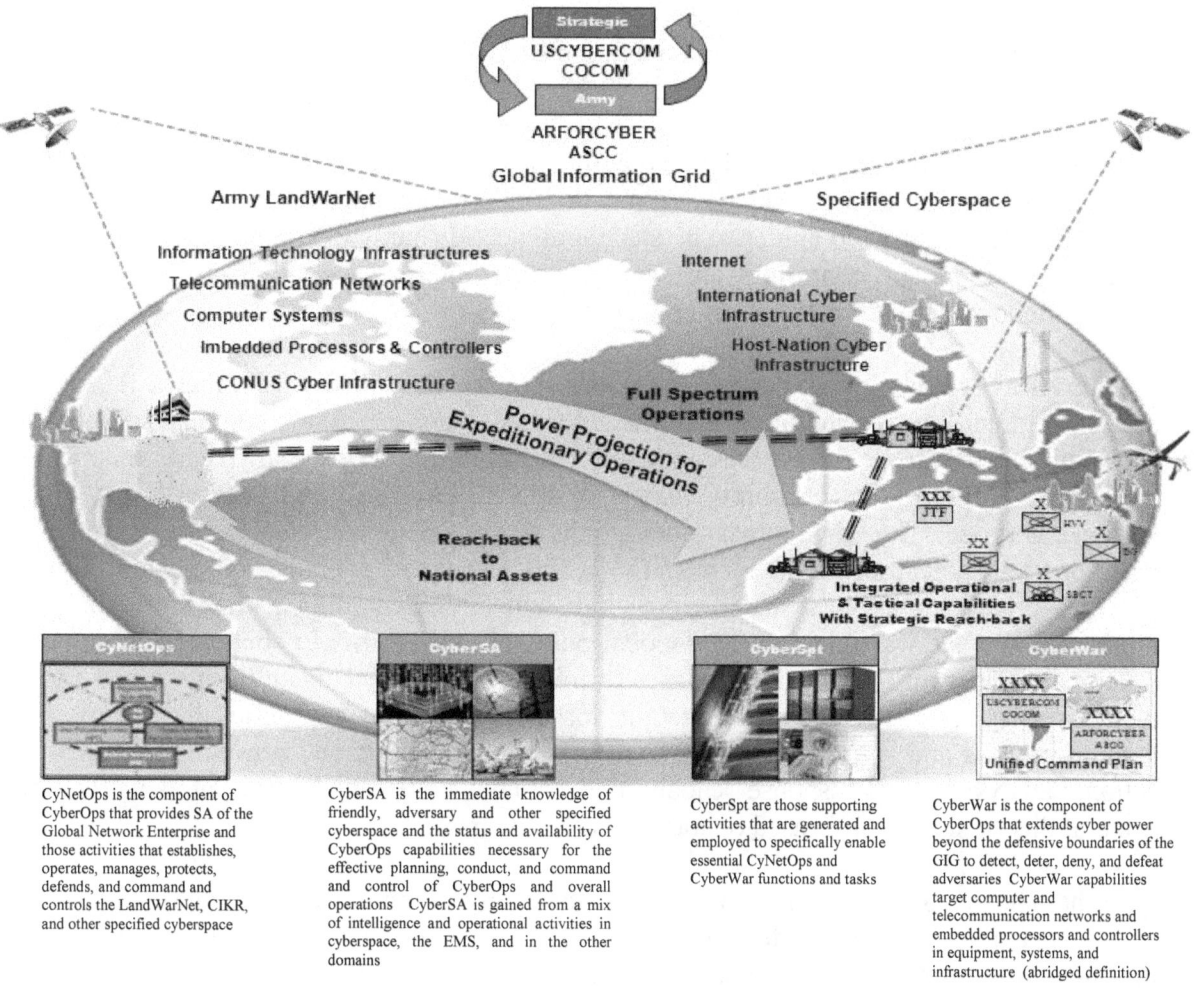

Figure B-1. Operational overview 1

b. For DOD, USCYBERCOM plans, coordinates, deconflicts, and conducts CyberOps. USCYBERCOM runs the Joint Interagency Coordination Group that executes interagency and key nation coordination and SA and information sharing. Operations are run out of its Joint Operations Center and Joint Intelligence Operations Center, both of which are tied to the Integrated Cyber Center, which coordinates U.S. government CyberOps.

c. CyberOps are conducted at the national, joint, and Army levels by both generating and operating forces. The Army provides trained, manned, and equipped Soldiers to USCYBERCOM, U.S. Northern Command (USNORCOM) U.S. Pacific Command (USPACOM), and to all combatant commanders and joint forces through the request for forces process. These forces are capable of conducting offensive and defensive CyberOps for joint operations and can support tactical operations through global reachback or support teams.

d. U.S. Northern Command (USNORTHCOM) and U.S. Pacific Command (USPACOM) are responsible for defense support to civil authorities (DSCA) for domestic emergencies, designated law enforcement, and other activities. DSCA tasks include vulnerability assessment, incident impact analysis, malware analysis, mitigation techniques, characterization, and digital media analysis capabilities. DSCA tasks do not include CNA and CNE, day-to-day law enforcement and intelligence support, and operation and defense of the GIG; however, these may be executed in defense of the homeland.

e. CYE capabilities are provided to expeditionary forces through USCYBERCOM global reachback support and through support teams in support of combatant commanders. Cyber attacks conducted against computer and telecommunication networks, computer systems, and embedded processors and controllers are executed under appropriate authorities through global reachback support, support teams, or organic capabilities.

f. The Army will continue to develop a global network enterprise (GNE) that will centralize control of the LandWarNet and other specified cyberspace under a single command. This GNE will provide LandWarNet unity of command by migrating loosely affiliated, independent networks into a true global capability that is designed, deployed, and managed as a single, integrated enterprise. The GNE cornerstones are the Army network service centers, consisting of network operations and security centers (NOSC) and associated computer emergency response teams, area processing centers, and regional hub nodes. NSCs align with the theater signal commands to provide warfighters with a global plug and play ability to connect to joint, Army, and commercial networks through all phases of joint operations. This ability to connect to the network enables greater freedom of action for Army forces throughout the Army force generation process and strengthens network defense through improved prevention, monitoring, detection, analysis, and response capabilities. The GNE enables joint and coalition interoperability to support increased operational effectiveness.

g. The Theater Signal Command is the senior Army communications headquarters in the theater. This mission includes telecommunications, engineering, installation, operations and maintenance responsibility for all deployed strategic signal forces in theater.

h. The Theater Network Operations and Security Center (TNOSC) supports the theater Army service component command (ASCC) by conducting 24/7 CyNetOps and security management on the theater information grid. The TNOSC plans, directs, and monitors CyNetOps and network defense; provides SA and reporting of theater information grid systems and networks; executes technical control and enforces compliance; and monitors and enables content management.

i. The regional computer emergency response team (RCERT) plans, synchronizes, and conducts CNO, provides technical and threat analysis, and directly supports the ground component commander, ASCC, and Theater Signal Command. The RCERT monitors the theater sensor grid, provides SA of threats and predictive threat analysis, conducts cyber incident response handling, and provides computer network defense (CND) and threat education for the units in theater.

j. The corps has a NOSC providing organic CyNetOps services. The NOSC synchronizes network management systems, including information assurance and CND, event correlation systems, and network monitoring systems. Currently, a corps does not have an assigned capability to plan and integrate all aspects of CyberOps.

k. The division has a NOSC providing organic CyNetOps services. These services include the coordination of all engineering, installation, operation, maintenance, management, and defense of the division information network. The NOSC is responsible for providing CyNetOps capabilities for distributed operations, battle command on the move, and in response to the commander's tactical requirements. However, the division must operate and defend its own network without augmentation from higher headquarters. Additionally, divisions do not have an assigned capability to plan and integrate all aspects of CyberOps.

l. Brigade and brigade combat teams (BCT) perform CyNetOps functions with its organic capabilities. Similar to the division, the brigade and BCT are required to operate and defend their own networks without augmentation from higher headquarters. This includes providing effective network management and network defense across all organic networks. In addition, the BCT provides the organic common services of messaging, collaboration, storage, and security to its subordinate elements. Currently, brigades and BCTs do not have an assigned capability to plan and integrate all aspects of CyberOps.

m. Battalions rely on their brigade or BCT for core services, network accessibility, and network defense. The battalion S-6 performs all the planning and operations associated with the main and tactical command posts, including establishing connectivity with adjacent, subordinate, and higher elements. Currently, battalions do not have an assigned capability to plan and integrate all aspects of CyberOps.

n. Companies rely on their battalion for core services, network accessibility, and network defense. The company performs all the planning and operations associated with the command post, including establishing connectivity with adjacent, subordinate, and higher elements

o. The total force. The Army National Guard (ARNG) and the Army Reserve (USAR) benefit from their associated civilian, academic, industry, and interagency communities to obtain Soldiers with specialized CyberOps skills, capabilities, and experience. The active component readily leverages the capabilities of the reserve component because they provide expanded capacity in areas that are often too expensive and too time consuming to handle alone. An example of this is malware reengineering. Reserve component forces also support Army organizations at the National Security Agency sites in Maryland, Georgia, Hawaii, Texas, and Colorado. Both the active and reserve components benefit when respective units share a habitual relationship in developing specific capabilities for teams, organizations, or force structure.

(1) The ARNG. A key component of the Army's total force CyberOps capability, the ARNG maintains CyberOps capabilities in the 54 joint force headquarters–state supporting both the Army and the states under their Title 10 and Title 32 authorities, respectively. Their priority is to establish and maintain a secure cyber environment for their state through CyNetOps by protecting critical cyberspace nodes, developing CyberSA, and providing support to civil authorities for incident response and protection of critical infrastructure. The ARNG is the Army's expert for protecting CIKR. It supports the Army and USCYBERCOM with CyNetOps, CyberSpt, and limited CyberWar capabilities.

(2) The USAR. The USAR provides trained and ready personnel to perform CyberOps in support of joint, Army, and combatant commander mission requirements. These personnel bring a maturity and depth of experience providing ready operational support to current operations. Uniquely, the USAR will have a direct link to U.S. Strategic Command contingency plans, allowing them to mobilize personnel to support ARFORCYBER plans and operations that support USCYBERCOM CyberOps. The USAR is expeditionary in nature and supports Army, ARFORCYBER, and USCYBERCOM with CyNetOps, CyberSpt, and limited CyberWar capabilities.

Appendix C
Operational Vignettes

C-1. Vignette context
This chapter uses a series of vignettes to illustrate the Army's operational framework, introduced in chapter 4, for conducting CyberOps as part of FSO from peacetime engagements to global war. The vignettes are consistent with the general framework of the multi-level security-1 scenario. Each vignette will be associated generally with one or more of the phases of the Joint Phasing Model from Joint Publication 3-0 (figure C-1) to get at specific actions and required capabilities from the strategic down to the tactical level. These implications and required capabilities are described in greater detail in appendices D and E and a consequent integrated list of preliminary DOTMLPF questions in appendix F.

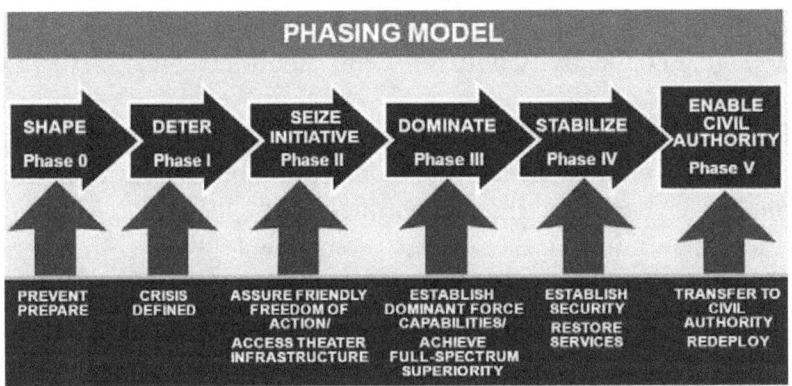

Figure C-1. The joint operations six phase model

C-2. Prephase 0 operations

a. This vignette focuses on normal, day-to-day operations and a typical Army support mission for a stability and support operation as part of a combatant command's theater security cooperation plans (TSCP). In the course of enduring peacetime operations, the theater combatant command in forward bases, and with supporting elements in the continental U.S. (CONUS) and outside the CONUS, must operate and defend the network, its associated systems, and the data resident on or in transit across the network. This includes establishing the public-private partnerships required to secure the commercial segments that are "inside" the Army's or other specified cyberspace. These operations occur while undergoing nearly continuous attacks and penetration attempts from a variety of threats ranging from non-state to state actors at all levels of sophistication, as well as insider threats and nonmalicious actors. The combatant command may also deploy forces for a variety of operational activities from nation building and counternarcotics activities to foreign internal defense and non-combatant evacuation operations.

b. Each of these activities requires a level of cyberspace operations support to gather intelligence on potential threats in the area of operations, extend, operate, and defend networks and services, to provide necessary command and control support. Combatant commands and subordinate echelons conduct intelligence assessments to identify commander's cyber intelligence requirements. Joint and Army elements conduct cyber exploitation to answer and update these assessments and requirements.

C-3. Vignette 1: phase 0 through phase 1, shape and deter

a. The vignette continues through the initial onset of a crisis and the actions taken to shape the operational environment and deter the adversary (figure C-2). This vignette will describe the Army CyberOps operations and capabilities required to effectively support the combatant commander's mission.

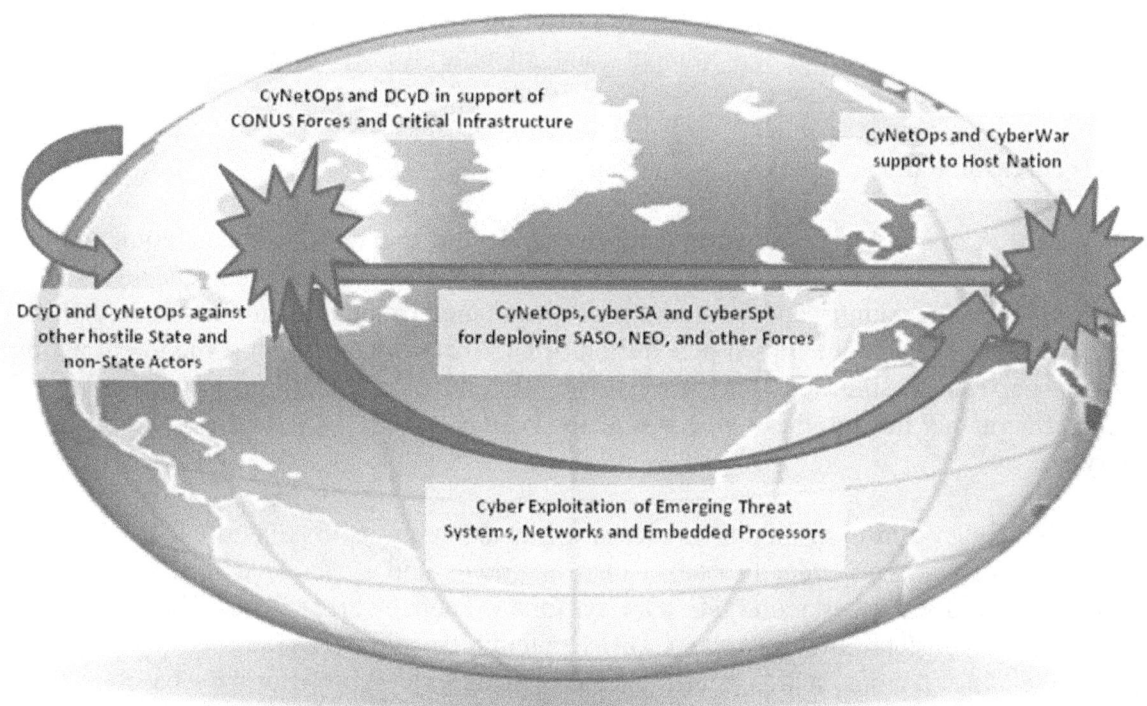

Figure C-2. Vignette 1: phases 0 & 1 – shape and deter

b. Situation.

(1) Friendly forces are conducting normal peacetime training operations, including leader development, education and training, RDT&E, and other activities to shape the OE. There is no traditional phase 0 or peacetime in cyberspace as adversaries continuously seek to conduct cyberspace operations, particularly exploitation, against the U.S. and its allies in order to pursue

their strategic objectives. Nation-states and other adversaries attempt to penetrate friendly networks to gain an information advantage and focus their cyberspace capabilities on collecting information about friendly interests, plans, technical data, and intentions.

(2) In this vignette, a large technologic peer-competitor and other adversaries also seek to undermine U.S. and coalition efforts to support a host nation that has requested assistance in providing stability and support during actions by a hostile neighboring country. These initial threat activities include attempted physical and cyber penetration and attempted disruption of host nation water and electrical services. As a crisis emerges, and an additional threat is added within the combatant commander's area of responsibility, the resources used to conduct day-to-day enduring operations are still required. Additional resources are likely to be required to provide support for each crisis as it emerges, surging forces and resources based on the commander's requirements, priorities, and intent.

c. Mission. The theater combatant command will continue to conduct normal peacetime operations. The commander's intent is to execute the TSCP to defeat, dissuade, and deter threats against the host nation and U.S. interests. The ASCC has been directed to conduct stability and support operations in order to restore essential host nation services and support civil security to provide a secure environment. This includes meeting the critical needs of the populace, gaining support for the host nation government, and shaping the environment for unified action, coalition, and host nation success.

d. Threat actions.

(1) A state, peer competitor nation outside the theater combatant command area of operation will not directly challenge U.S. or coalition forces but will rather tacitly support the hostile neighboring country and insurgent activities to undermine U.S. and coalition efforts. This state will likely escalate attempted penetration of U.S. networks to support the threat nation with intelligence. The hostile neighboring country will conduct overt efforts to undermine the host nation, but avoid raising these operations to a level where U.S. and coalition forces will intervene.

(2) Cyber efforts by all adversaries will include attempts to penetrate U.S., coalition, and host nation networks in order to collect data on forces and systems, with a special emphasis placed on military networks and SCADA systems. Their efforts will be directed against any deployed U.S. or coalition forces as well as potentially against U.S. based rapid deployment forces, their installations, and supporting CIKR. This exploitation forms a baseline for future determination of U.S., host nation, and coalition capabilities and intentions.

e. CyberOps support for the mission. Through the combatant command ASCC and ARFORCYBER, joint and Army CyberOps capabilities will continuously support the operating and generating forces, as well as the deployed forces conducting the combatant command's TSCP and stability and support missions in theater. This section will describe how the ASCC, ARFORCYBER, and subordinate and supporting forces, as well as other National and joint organizations and forces will employ specific CyberOps capabilities to support achieving the combatant commander's mission and intent. This will be done within the framework of

CyNetOps, CyberWar, CyberSpt, and CyberSA to demonstrate how these efforts enable commanders to effectively conduct FSO. The following actions occur at each echelon:

(1) Combatant command. At this level, cyber ROE are established in conjunction with the President, the Office of the Secretary of Defense (SECDEF), the Joint Staff, and partners within the U.S. intelligence community. Strategic plans are reviewed, adjusted as needed, and campaign design activities occur, to include the development of engagement strategies in which cyberspace operations may play a key role in content delivery to selected audiences.

(2) USCYBERCOM. Resource allocation is closely examined to ensure adequate support for the combatant command and the emerging threat in theater, as well as taking steps to surge enough resources to mitigate the increasing threat to both deployed forces and the CONUS base. Partnerships and command relationships across the Army as well as with Federal, state, and local law enforcement and the Department of Homeland Security begin to focus on mitigation of this new threat, while remaining vigilant to the day-to-day threat.

(3) Theater ASCC. Conducts detailed mission analysis to identify Army cyber force and resource requirements needed to support the theater combatant commander's mission.

(4) ARFORCYBER. Army cyber forces and resources are given orders to prepare for surging to support the theater combatant command and theater ASCC.

(5) The joint task force (JTF) may be designated to begin detailed crisis planning and prepare to receive forces and resources and may prepare to conduct reception, staging, onward movement, and integration activities if directed. Affected echelons prepare to receive cyber planners and integrators.

f. CyNetOps

(1) The Army operating and generating forces continuously conduct CyNetOps to operate and defend LandWarNet and support unit command and control activities. CyNetOps are conducted at the joint level through the brigade and BCT to the battalion and potentially company levels (reference appendix B) to ensure network services and the confidentiality, integrity, and availability of information on these systems. Deploying units will connect to the network service center for global network connectivity, enterprise services, and network management to support all aspects of FSO. Assigned and attached cyber planners at echelons from BCT (or lower if required) through combatant commands plan, prepare, execute and assess the effectiveness of CyberOps as part of unified action. This ensures access to LandWarNet, required battle command and control systems, the EMS, supporting critical infrastructure, and other key capabilities. This necessitates complete integration, involvement, and effective partnerships with Army active and reserve components, the generating and operating forces, and public-private partnerships.

(2) All echelons of the command continuously conduct defensive actions to protect themselves and the critical infrastructure on which they depend from cyber, electronic, and directed energy attacks. They are also prepared to mitigate the impact of and fight through an

attack, and operate effectively under degraded conditions. This includes defensive aspects in CONUS, in theater, and globally. Effective public-private and law enforcement and counter intelligence partnerships and activities, take proactive actions, share information, actively analyze and help mitigate the threat on a daily basis. As part of the combatant command's mission, this includes the necessary CyberOps support to the host nation.

g. CyberWar.

(1) Army and joint forces at the USCYBERCOM, ARFORCYBER, JTF, and potentially BCT echelons conduct CyberWar activities in phases 0 and 1. CyA and CyE are used to destroy, deny, degrade, disrupt, and deceive adversary data, computers, systems, embedded processors and controllers and thereby reducing the effectiveness of adversary decisionmakers. Unit cyber planners, in partnership with the national intelligence and law enforcement community coordinate and synchronize the conduct of operations to access and exploit adversary systems and networks to build friendly CyberSA and support the commanders' efforts to understand threat capabilities, vulnerabilities, plans, and intentions. This is a small but vital part of the commander's overall SA. CyberWar efforts are also used to gain access to the requisite portions of cyberspace used to support information activities, psychological operations product dissemination, or in direct support of isolating or disrupting adversary command and control just prior to and during tactical direct action operations.

(2) Following friendly tactical operations, exploitation may be employed to gather adversary target or battlefield damage assessment information. To effectively accomplish this, the combatant command develops recurring physical and remote access to adversaries' hardware and software, as well as friendly knowledge management systems that enable the ability to aggregate, manage, decrypt, linguistically translate, analyze, and report on all data collected to the supported unit commanders at all echelons and, when appropriate, to the host nation.

h. CyberSpt.

(1) CyberSpt activities are conducted in support of normal peacetime operations and the combatant command's mission. Deployed elements conducting stability and support operations in and around the host nation conduct site exploitation activities, to include cyber aspects of the exploitation. This exploitation may result in detailed forensic exploitation, reverse engineering and analysis of threat data, systems, and tactics, techniques, and procedures (TTP) to continue to gain and protect the advantage. Media pulled from the field is physically and/or virtually sent to higher echelons in theater or CONUS as needed for more detailed analysis. When adversaries or other actors in cyberspace attempt to attack or exploit friendly or supported host nation data, systems, or networks, units use organic capabilities to perform incident response handling activities associated with a suspected incident to learn and evolve from that incident. Lessons learned are then incorporated into an ongoing program of RDT&E, vulnerability assessment and mitigation, penetration testing, and leader and Soldier training programs to continue to maintain a relative information advantage over the adversary while supporting the host nation and the commander's intent in the theater.

(2) Policy and legal advice on cyber issues is provided to commanders at every echelon in the command to support awareness, understanding and implementation through all phases of conflict in the CONUS base and deployed. This is translated into appropriate ROE that address appropriate use of friendly, adversary, and other specified cyberspace.

i. CyberSA. CyberSA is derived from a detailed understanding of friendly, adversary, and other specified cyberspace. CyberSA is a component of overall SA and is only presented separately here to support greater understanding and awareness of this new element to that SA. Key warfighting functions at combatant command through battalion level, including intelligence, command and control, and movement and maneuver play important roles in contributing to CyberSA, each updating their portion of the commander's common operating picture (COP). CyNetOps personnel, enabled by sensors and other capabilities that detect, aggregate and report on the operation and health of systems, networks, and the associated content. Intelligence representatives focus their attention on the adversary and relevant cyberspace in order to effectively support lethal and nonlethal operations and intelligence activities. The staff cyber planner and integrator, empowered by effective data visualization capabilities supports the fusion of all CyberSA into a single, coherent picture to ultimately support the commander's decision-making process.

C-4. Vignette 2: Phases 1 and 2, force deployment

a. This vignette focuses on the strategic and operational CyberOps capabilities required for force deployment, initial introduction of forces into theater, and simultaneous civil support operations (figure C-3)..

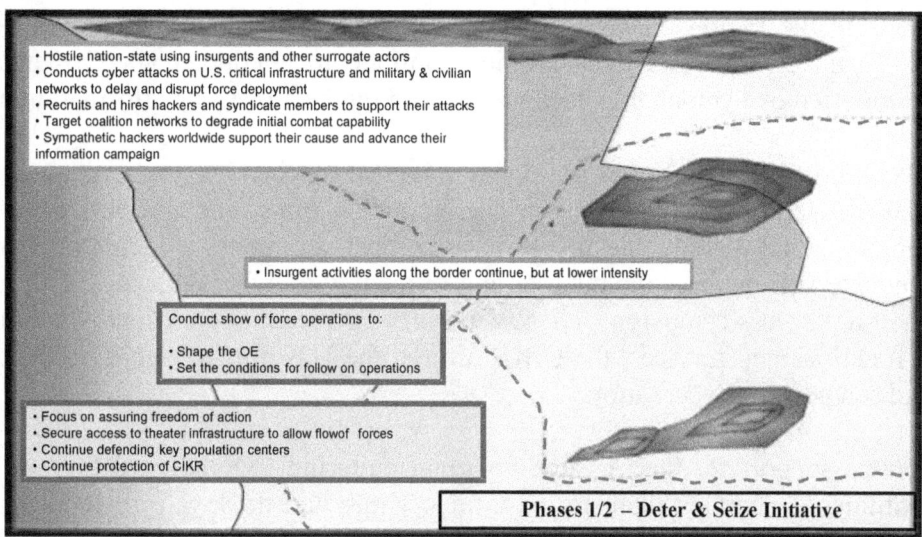

Figure C-3. Vignette 2: Phases 1 and 2 – Deter and Seize Initiative

b. Situation. A friendly host nation has requested U.S. and United Nations assistance to deter and defend them from a neighboring hostile nation-state. This nation-state is using insurgents and the actions of other surrogate actions for the advancement of their own personal and criminal objectives. The U.S. has limited forces initially in theater to conducting liaison and training with host nation forces and government agencies. Additional forces are being introduced to theater to deter and enable seizing the initiative from hostile actors. Adversary forces conduct cyber attacks on U.S. critical infrastructure and military and civilian networks to delay and disrupt force deployment. The SECDEF has tasked the DOD to conduct civil support operations to protect both physical and cyber critical infrastructure.

c. Mission. U.S. and coalition forces deploy into theater to support the requesting friendly nation. Friendly forces will conduct show of force and distributed operations to secure the aerial and sea ports of debarkation and key terrain and to deter further aggression of the neighboring hostile nation-state. Joint forces will also conduct civil support operations in CONUS to protect both physical and cyber critical infrastructure. CyberOps activities conducted in phase 0 continue. The primary focus of CyberOps during phases 1 and 2 is on assuring freedom of action and access to theater infrastructure so follow-on forces can seamlessly flow into the area of operations. Coalition forces engage in simultaneous offensive, defensive, and stability operations to shape the OE and set the conditions for decisive operations. These include defending key populations and CIKR, and continuing civil support operations.

d. Threat actions.

(1) The neighboring hostile nation-state will conduct cyber attacks on U.S. civilian and military targets to delay and disrupt force deployments and require U.S. and coalition nations to divert CyberOps and other resources for civil support operations. The adversaries will recruit and hire hackers and syndicate members to attack U.S. and coalition nation targets and inspires independent hackers to do the same. These cyber attacks target U.S. CIKR, A-Land CIKR, and U.S. and coalition interests worldwide. Adversaries will also leverage cyberspace to further its information strategy to influence world opinion against U.S. intervention, engender support of the populations in the region, inspire, and motivate their own populace, and foster anti-U.S. sentiment. The adversaries and their proxies continue to conduct CyberOps against targets worldwide, supported by sympathetic hackers inspired by the adversary information campaign.

(2) With hostilities appearing imminent, adversary CyberOps and EW attacks will target coalition networks, computer applications, and vehicles, ships, and aircraft to degrade initial combat capability. This will include attacks in the global commons, to include cyberspace. The hostile nation-state will increase the number and frequency of cyber attacks and direct actions in support of their antiaccess campaign. These actions are directed at the friendly host nation's key government functions and services, the CIKR around the U.S. forces identified for deployment, and aerial and sea ports of debarkation.

e. CyberOps support for the mission. Normal enduring CONUS and forward base cyber activities continue. CyberOps planners now must surge and deploy from their assigned units through ARFORCYBER and the theater ASCC to their supported units at echelons from the theater combatant commands to forces possibly as low as the BCT based on JTF task organization and priorities. These planners will augment and support all aspects of the commander's deployment planning and execution cycle. CyberOps will provide CyberSA and overall SA for the commander. CyNetOps capabilities will be critical for seamless force deployment and immediate employment upon arrival in theater. CyberWar capabilities will enable an operational advantage. CyberSpt capabilities will be tailored to the commander's mission to provide responsive support to both CyNetOps and CyberWar activities. The following actions occur at each echelon:

(1) Combatant command. At this level, cyber ROE are modified and the echelons at which decisions about actions in cyberspace, including the redissemination of approved engagement products may be pushed down to the JTF commander. The combatant command coordinates with USCYBERCOM and the national intelligence community to conduct cyber exploitation of specific threat targets, links, and nodes in support of strategic and operational targeting, increasingly focused CyberSA, and to support the execution of the campaign.

(2) USCYBERCOM surges to support precombat deployment activities and entry operations. Additional forces are requested from ARFORCYBER to further augment planners, network operators, and defenders, as well as additional resources to support CyA and CyE activities in support of the theater combatant command.

(3) Theater ASCC. Identifies additional resource requirements needed to support the theater combatant command mission and passes requirements to the theater combatant command for forwarding as part of the request for forces process.

(4) ARFORCYBER Army cyber forces and resources are given orders to prepare for surging to support the theater combatant command, theater ASCC. The Network Enterprise Technology Command continues to operate and defend the network, critical infrastructure, and key resources.

(5) JTF through BCT may be designated to begin detailed crisis planning and prepare to receive forces and resources and may prepare to conduct reception, staging, onward movement, and integration activities if directed.

f. CyNetOps

(1) While CyNetOps continues to gain advantage, CyNetOps in these phases protect the advantage that the global network already provides. The GNE provides for global connectivity which includes a collaborative and en route planning capability. This includes a seamless connection to the global network upon arrival in theater and for initial distributed and economy of force efforts. Division, brigade, and BCT elements deploy into theater with self-contained networking capabilities that plug into the network service center for global connectivity. Division, brigade, and BCT elements perform CyNetOps functions with their organic capabilities and are required to operate and defend their own networks without augmentation from higher headquarters. This includes providing effective network management and network defense across all organic networks.

(2) In addition, the BCT provides the organic common services of messaging, collaboration, storage, and security to its subordinate elements. Currently, brigades and BCTs do not have a designated capability to plan and integrate all aspects of CyberOps. CyNetOps employs protective measures that ensure the availability of critical battle command and control systems that provide the commanders a means for making more informed decisions to achieve

objectives on the ground. CyNetOps, when directed, also provides CyNetOps support to host nation and coalition forces and, with joint force headquarters support, augments these organizations with expertise and capabilities.

(3) CyberSA becomes more dynamic and complex due to the nature of operations during phase 2 and leading into phase 3. CyNetOps capabilities must defend against both cyber and EAs to include directed energy attacks. Understanding the congested and contested EMS is critical for the commander's CyberSA to identify areas where there is the potential for the degradation of key capabilities. Civilian and military applications in the RF spectrum may lead to degraded CyberOps capabilities in certain portions of the OE.

(4) Army active and reserve component CyberOps personnel will support civil support operations. This will be done under specific authorities and normally would include CyNetOps and CyberWar personnel conducting incident impact analysis, mitigation techniques, threat characterization, vulnerability assessment, malware analysis, and digital media analysis. If the civil support operations turn into a homeland defense mission, then all four components of CyberOps could be used to support the mission.

g. CyberWar.

(1) CyberWar capabilities will initially focus on developing CyberSA and conducting CyE and DCyD. These activities would progress to CyA activities to deny the adversary's ability to disrupt the commander's plans and in support of direct operations to kill or capture identified targets. CyberWar capabilities will be a combination of organic and support element joint and Army capabilities. While many global capabilities will be available, the Army will provide the commander with close access capabilities where physical proximity is required. Organic Army staff cyber planners and integrators will provide the commander the expertise to integrate CyberWar capabilities into FSO.

(2) Joint and Army support elements are provided to augment organic division, brigade, and BCT CyberOps integration and planning capabilities to ensure available CyberOps resources are integrated within the overall operational plan and to monitor their execution to ensure they achieve the commander's intended mission effects. CyberWar planning activities may include "preplanned, on call" priority missions as part of the overall concept of operations. Battalion and company staffs have trained personnel for CyberOps planning and integration. DCyD will integrate with CyNetOps for the commander's defense in depth. A range of CyberWar capabilities will enable cyber threat tipping and cueing, counter reconnaissance, and counterintelligence efforts.

h. CyberSpt. Robust RDT&E and timely, responsive acquisition processes allow the Army to stay apace with technologic advancements and adversary adaptations of existing technologies. Partnering with internal, Federal, state, and local law enforcement agencies help to support the pursuit and investigation of criminal organizations and independent hackers working against friendly forces. Penetration testing and red, blue, green teams are used for vulnerability and security testing. CyberSpt activities focus on the cyber aspects of site exploitation, forensics, mitigation, and remediation are used to gain and protect advantage. Prompt site exploitation protects advantage by identifying threat TTP so countermeasures can be developed during expeditionary operations. The unit legal officer provides the commander an assessment of all CyberWar targets and actions within the confines of the ROE and all relevant U.S., host nation, and international laws.

i. CyberSA. As described in Vignette 1, CyberSA is derived from a detailed understanding of friendly, adversarial, and other specified cyberspace. The challenge to CyberSA in phases 1 and 2 is CyNetOps is very dynamic during force deployment and CyberWar capabilities take time to fully develop an understanding of adversarial and other specified cyberspace. Close coordination and partnership between the operating and generating forces, public-private stakeholders, and Army and unified action and multinational partners will be critical to rapidly develop and maintain CyberSA. The staff cyber planner and integrator will be responsible for the fusion of friendly, adversarial, and other specified cyberspace SA into a single, coherent picture to ultimately support the commander's decisionmaking process.

C-5. Vignette 3: Phases 2-4, major combat operations (MCO)

a. The focus of this vignette is on the operational and tactical CyberOps capabilities required to support MCO with simultaneous stability operations (figure C-4).

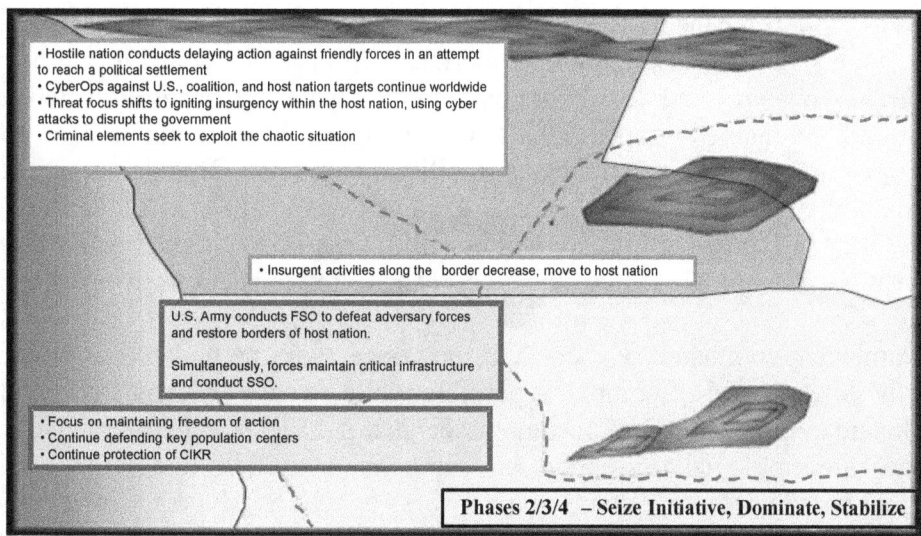

Figure C-4. Vignette 3: phases 2, 3, and 4 – seize initiative, dominate, and stabilize

b. Situation. The hostile neighboring nation-state remains undeterred and has taken actions resulting in the joint force conducting offensive operations. The noncontiguous battlefield also results in simultaneous stability operations in certain portions of the host nation. Civil support operations have been successful in the U.S. and the sole focus for the U.S. is on MCO and stability operations.

c. Mission. The Army, in concert with its unified action and coalition partners, will conduct FSO to defeat the military and other adversary forces of the neighboring hostile nation-state that are occupying portions of the host nation.

d. Threat actions.

(1) The neighboring nation-state military and other adversary forces will conduct a delaying action against U.S., coalition, and host nation forces in an attempt to reach a political settlement in line with its objectives. Adversaries will conduct CyberOps against targets worldwide, supported by sympathetic hackers inspired by the hostile force's information campaign. With hostilities appearing imminent, adversary CyberOps and EW attacks will target coalition networks, computer applications, and vehicles, ships, and aircraft to degrade initial combat capability. Cyber attacks from the neighboring hostile nation-state forces may decrease, but attacks from criminal and outside elements (hacker unions and others rallying to the adversary cause) will likely increase.

(2) Once hostilities begin, the threat focus shifts to igniting an insurgency against the host nation's government, using cyber attacks to disrupt the government. This includes using CyberOps capabilities for command and control, fundraising, recruiting, and otherwise fostering an insurgent environment, supporting, and recruiting criminal activity in cyberspace against banking and commercial sites as criminal elements seek to exploit the current chaotic situation in the host nation, and continuing the use of CyberOps capabilities to support their overall information strategy.

e. CyberOps support for the mission. Joint and Army CyberOps capabilities will support all aspects of the commander's plan for FSO. CyberOps will provide CyberSA and overall SA for the commander in a dynamic and distributed battlefield. CyNetOps capabilities will be dynamic due to distributed, simultaneous operations that accompany battle command on the move in addition to the requirements for simultaneous stability operations. Organic and support element CyberWar capabilities will be used to gain and protect advantage and place adversaries at a disadvantage. Responsive CyberSpt capabilities will be required to adapt to a thinking enemy that tailors advanced technologies to its objectives. The following actions occur at each echelon.

(1) Combatant command. Cyber ROE may once again be updated to provide as much flexibility to the lowest level of command possible while remaining cognizant of the potential strategic implications of cyber operations and associated intelligence gain-loss issues. Strategic cyber resources are employed to support air campaign activities, disruption of threat early warning, air, and missile defense systems and networks as well as adversary information systems platforms in close coordination with early entry special operations and ground force operations.

(2) USCYBERCOM. The main effort may shift from exploitation to attack of adversary systems, links, nodes, and data.

(3) Theater ASCC. Continues to support combatant command requirements as they are identified.

(4) ARFORCYBER provides priority of support to the theater ASCC as new requirements emerge. The Network Enterprise Technology Command continues to operate and defend CONUS systems and networks while providing priority of support to the theater combatant commander.

(5) JTF through BCT continue to leverage cyber planners and integrators as key members of their battle staffs working with intelligence, fires, and information engagement cells to plan, coordinate, and synchronize cyber capabilities in support of FSO and ensuring cyber SA is an accurate, relevant component of the commander's COP.

f. CyNetOps

(1) CyNetOps during major combat operations and stability operations will need to be flexible and agile. CyNetOps capabilities will be provided from the joint to the battalion and possibly lower echelons (see appendix D and E). Protecting the CyNetOps advantage is critical in phases 2, 3, and 4 because of the critical battle command and other capabilities that rely on CyNetOps capabilities. U.S. and coalition forces use CyNetOps capabilities to enable battle command on the move, distributed operations, and stability operations, while preparing to operate in a degraded CyberOps condition. Assigned and attached cyber planners and integrators at echelons from battalion through the joint force will plan, prepare, execute, and assess the effectiveness of CyberOps in support of MCO and stability operations.

(2) During phase 4, CyNetOps enhancements will result in a more robust infrastructure (for example, commercialization) freeing up forces for other tasks and preparing for the transition to phase 5. CyberSA becomes more dynamic and complex due to the nature of operations during phases 2, 3, and 4.

g. CyberWar.

(1) CyberWar and enabling capabilities will exploit and attack computer and telecommunication networks and embedded processors and controllers in equipment, systems, and infrastructure, in accordance with appropriate authorities in support of the commander's objectives. This is the first time that CyberWar attacks on tactical target embedded processors and controllers in equipment, systems, and infrastructure are conducted to disadvantage the adversary. During maneuvering, a different blend of CyberWar and enabling capabilities will be employed due the responsiveness and authorities under which each is conducted.

(2) Unit cyber planners and integrators, in partnership with the national intelligence and law enforcement community, coordinate and synchronize the conduct of operations to access and exploit adversary systems and networks. Following friendly tactical operations, exploitation may be employed to gather adversary target or battlefield damage assessment information. To effectively accomplish this, recurring physical and remote access to adversaries' hardware and software is required. DCyD will be emphasized in phases 2, 3, and 4 due to the fluid nature of the operations and due to the dynamic nature of CyNetOps, defense in depth will be absolutely critical.

h. CyberSpt. Cyber aspects of site exploitation, forensics, mitigation and remediation, and responsive RDT&E capabilities are the initial CyberSpt capabilities that support MCO and stability operations. As stability operations become more prevalent, vulnerability and security assessments will increase in importance.

i. CyberSA. CyberSA will be a challenge to establish and maintain during the dynamic operations of phases 2, 3, and 4, but they are absolutely critical so the commander can make informed decisions. The simultaneous execution of phases 2, 3, and 4 also dictates that adequate analysis and understanding is developed as the majority of the force transitions to phase 4 and prepares for phase 5.

C-6. Summary

This appendix used a series of vignettes to describe how CyberOps can be integrated to support FSO. Appendix D and E capture the required capabilities that were described in these vignettes. These required capabilities will serve as the starting point for the CyberOps CBA.

Appendix D
Required Capabilities

D-1. Introduction

a. This Appendix lists the capability statements that have emerged from the integrated concept development team's (ICDT's) development of this pamphlet. The capability statements within this appendix represent the initial identification of capabilities needed for CyberOps, based on the analysis performed to date by the ICDT. This body of work represents a point for departure for follow-on analysis.

b. Specifically, this appendix provides the Army's future force with broad operational CyberOps capability requirements with which to use as a starting point for the CyberOps CBA. It is the culmination of experiments, symposiums, conferences, working groups, literature reviews of joint and Army concepts and doctrine, joint and higher direction and policy guidance, and subject matter expert (SME) input and insights that have led to the central idea, framework, and vignettes in this pamphlet. The ICDT derived these required capabilities from joint and Army concepts, JCAs, the results of Unified Quest (UQ) 2009 (UQ 09), Omni Fusion (OF) 2009 (OF 09), the 2009 Space and Cyberspace Seminar Wargame (UQ 09 SWG), quadrennial roles and missions report (QRM), battle command essential capability (BCEC) documents[15], and other venues and documents. Operational SMEs from within the information and cyberspace ICDT and those leveraged by the ICDT during this pamphlet development contributed to the required capabilities in this appendix. These required capabilities are broken down into specific tasks, conditions, and standards during the follow on CyberOps CBA. Required enabling capabilities including EW and EMSO will be analyzed during the CBA by leveraging the specific Army concept framework, CBA, and JCIDS documents.

c. Required capabilities have four basic elements: organization (who), main idea (what), environment, parameters, and conditions (where and when), and reason (why). The future force is the organization for these required capabilities. Tables D-2, D-4, D-6, and D-8 in the next section further define the "who" by identifying the echelon at which the capability is required. If

---

[15] BCEC are the essential set of capabilities required by commanders to perform battle command during FSO. These TRADOC-approved BCEC are the foundational set of capabilities required by commanders and leaders at brigade, battalion, and company and below but are also fully applicable to the division, corps, and theater Army levels.

there was not a consensus for the echelon where a capability should reside, the ICDT used the lower echelon to ensure the proper analysis will be done in the CBA. As described in chapter 2 and appendix C, CyberOps are continuous so these capabilities are required across the breadth of the future OE and in all mission areas of FSO. The CBA will define specific tasks, conditions, and standards that support each capability.

D-2. DOTMLPF required capabilities

a. The applicable DOTMLPF domains are identified by their first letter in the third column of tables D-1, D-3, D-5, and D-7 for each capability requirement. In addition, the source of the requirement is noted in the last column.

Note: Appendix F contains an integrated DOTMLPF question list that will be used in future capability development efforts.

b. CyNetOps. To conduct CyNetOps effectively, the future force requires the capabilities listed in table D-1 below. These are only the higher level required capabilities as the CBA will reference the LandWarNet CONOPS and network transport required capabilities chapters for additional detailed requirements. CyNetOps supports the warfighting function and battle command critical systems and processes so accurate conditions and standards for the CBA tasks will be essential.

Table D-1
Required CyNetOps capabilities

#	Required CyNetOps Capability	DOTMLPF	Source / References
1	Operate an enterprise network capability for computer and telecommunication networks at multiple levels of classification, from Unclassified through Top Secret for brigade and above and Secret for below brigade. Including coalition and alliance classifications that includes the Army active and reserve components as well as the generating and operating force networks in order to provide end-to-end assured CyNetOps support to critical battle command and other capabilities. **Note**: This capability encompasses and supports the BCEC capability of, "A robust network capability: the force must possess a commander centric secure integrated and adaptable communications network consisting of line-of-sight and beyond-line-of-sight means."	O,T,M,P,F	global network enterprise construct (GNEC) TRADOC Pam 525-7-8 SMEs BCEC
2	Provide global connectivity to an enterprise communications network infrastructure in order to provide end-to-end assured CyNetOps support to critical battle command and other capabilities and ensure freedom of action. **Note**: This capability encompasses and supports the BCEC capability of, "A robust network capability: the force must possess a commander centric secure integrated and adaptable communications network consisting of line-	O,T,M,P,F	GIG 2.0 initial capabilities document (ICD) BCEC

#	Required CyNetOps Capability	DOTMLPF	Source / References
	of-sight and beyond-line-of-sight means."		
3	Provide enterprise communications networks which are interoperable with joint, Army, interagency, and multinational organizations to include nongovernmental organizations in order to provide end-to-end assured CyNetOps support to critical battle command and other capabilities and ensure freedom of action. **Note**: This capability encompasses and supports the BCEC capability of, "Joint, interagency, intergovernmental, multinational (JIIM) interoperability: The Army's battle command system must be able to exchange relevant operational information with JIIM partners, nongovernmental organizations and contractors."	O,T,M,P,F	GIG 2.0 ICD BCEC
4	Integrate coalition partner and other specified networks during garrison and deployed operations, including the capability to integrate into the networks of coalition partners with different intelligence sharing relationships in order to enable effective joint and/or multinational operations and ensure freedom of action. **Note**: This capability encompasses and supports the BCEC capability of, "JIIM Interoperability: The Army's battle command system must be able to exchange relevant operational information with JIIM partners, nongovernmental organizations, and contractors."	D,O,T,M,L,P,F	UQ 09 GIG 2.0 BCEC
5	Define the public-private partner roles, responsibilities, and authorities necessary in order to secure the commercial segments that are used by the Army or other specified cyberspace and ensure freedom of action.	D,O,T,M,L,P,F	TRADOC Pam 525-7-8
6	Provide a command and control system capability to obtain, process, and disseminate CyberOps information in order to facilitate the commander's decisionmaking and enable effective operations. **Note**: This capability encompasses and supports the BCEC capability of, "Execute a Running Estimate: The Army's battle command system must be able to continuously gather and track information to support tactical decisionmaking by providing a continuous assessment of current and future operations."	D,O,T,M,L,P,F	ARCIC BCEC
7	Provide globally authenticated users access to CyberOps capabilities in order to support distributed, remote, and battle command on-the-move operations and ensure freedom of action. **Note**: This capability encompasses and supports the BCEC capabilities of, "Execute tactical network	D,O,T,M,L,P,F	TRADOC Pam 525-7-8 GIG 2.0 ICD BCEC

#	Required CyNetOps Capability	DOTMLPF	Source / References
	operations: commanders need the ability to conduct tactical network operations (network management) and allocate network resources to maximize performance through all phases of an operation.," and "battle command on-the-move: The commander must have the ability to maintain situational awareness, make timely and informed decisions, and position himself at the decisive point during the battle."		
8	Aggregate all data regarding information systems resources connected to the network in order to provide end-to-end assured CyNetOps support to critical battle command and other capabilities and contribute to CyberSA.	O,T,M,L,P,F	GNEC
9	Monitor network and information system status and health; conduct system maintenance; and collect system demand history and usage rates in an entirely automated and real-time manner in order to provide end-to-end assured CyNetOps support to critical battle command and other capabilities and contribute to CyberSA.	D,O,T,M,L, P,F	TRADOC Pam 525-7-8UQ 09 SWG GNEC
10	Operate under a degraded cyber operations condition in order to conduct effective battle command and operations and to ensure freedom of action.	D,O,T,M,L, P	TRADOC Pam 525-7-8, QRM UQ 09 UQ 09 SWG OF 09
11	Provide defense-in-depth for LandWarNet and other specified cyberspace in order to provide end-to-end assured CyNetOps support to critical battle command and other capabilities.	D,O,T,M,L, P,F	TRADOC Pam 525-7-8 QRM
12	Protect against cyber and EA to include directed energy attacks in order to provide end-to-end assured CyNetOps support to critical battle command and other capabilities and ensure freedom of action.	D,O,T,M,L, P	UQ 09 UQ 09 SWG
13	Monitor for and report cyber threat events in real time in order to contribute to defense-in-depth and CyberSA.	D,O,T,M,L, P,F	TRADOC Pam 525-7-8 QRM OF 09 UQ 09 SWG GIG 2.0 ICD
14	Detect and monitor network intrusions and unauthorized activity in real time in order to contribute to defense-in-depth, provide end-to-end assured CyNetOps support to critical battle command and other capabilities, ensure freedom of action, and contribute to CyberSA.	D,O,T,M,L, P,F	TRADOC Pam 525-7-8, GNEC, GIG 2.0 ICD
15	Analyze and understand in real time the nature of	D,O,T,M,L,	TRADOC Pam

#	Required CyNetOps Capability	DOTMLPF	Source / References
	malicious and unauthorized activity occurring within the network in order to contribute to defense-in-depth, provide end-to-end assured CyNetOps support to critical battle command and other capabilities, ensure freedom of action, and contribute to CyberSA.	P,F	525-7-8, GNEC, GIG 2.0 ICD
16	Attribute actions on friendly, specified, and adversary networks in order to support CyNetOps and CyberWar actions.	D,O,T,M,L,P	QRM
17	Defend against and fight through a cyber threat event in order to conduct effective battle command and operations and ensure freedom of action.	D,O,T,M,L,P	TRADOC Pam 525-7-8, QRM, UQ 09 UQ 09 SWG OF 09
18	Provide physical and CyberOps protection against both lethal and nonlethal attacks on critical infrastructure and key resources during all phases of FSO in order to contribute to defense-in-depth, provide end-to-end assured CyNetOps support to critical battle command and other capabilities, and ensure freedom of action.	D,O,T,M,L,P,F	UQ 09 SWG
19	Gain awareness of, access to, and delivery of information, information services, and applications in order to provide end-to-end assured CyNetOps support to critical battle command and other capabilities.	D,O,T,M,L,P,F	TRADOC Pam 525-7-8 GIG 2.0 ICD
20	Support, enable, or provide CyberOps capabilities to host-nation or other actors in order to contribute to defense-in-depth, provide end-to-end assured CyNetOps support to critical battle command and other capabilities, and enable effective joint and/or multinational operations.	D,O,T,M,L,P,F	UQ 09 TRADOC Pam 525-7-8
21	Share information and collaborate with public and private partners in all aspects of network operations and CIKR protection in order to contribute to defense-in-depth and provide end-to-end assured CyNetOps support to critical battle command and other capabilities. **Note**: This capability encompasses and supports the BCEC capability of, "Enable Collaboration: Commander's and leaders need a common suite of collaborative tools to allow establishment of a collaborative environment to achieve shared understanding and ensure unity of effort."	D,O,T,M,L,P	UQ 09 SWG GIG 2.0 ICD BCEC
22	Ensure the availability, confidentiality, and integrity of essential CyberOps capabilities in order to provide end-to-end assured CyNetOps support to critical battle command and other capabilities.	D,O,T,M,L,P,F	QRM UQ 09
23	Develop a standard and sharable geospatial foundation in order to enable all battle command essential information	D,O,T,M,L,P,F	BCEC

#	Required CyNetOps Capability	DOTMLPF	Source / References
	requirements, create a common map foundation, and display and share this information on an interoperable COP.		
24	Create, change, and distribute mission orders (both voice and written) to include attached graphics in order to enable effective battle command communication between command posts, platforms, and leaders.	D,O,T,M,L, P,F	BCEC
25	Provide rehearsal and training support in order to prepare for operations using embedded rehearsal and training tools that accurately represent the spectrum of missions and environments.	D,O,T,M,L, P,F	BCEC

c. Table D-2 below lists the echelon at which each capability is required.

Table D-2
Required CyNetOps capabilities by echelon

Echelon/Capability #	01	02	03	04	05	06	07	08	09	10
Joint, combatant command	X	X	X	X	X	X	X	X	X	X
ARFORCYBER	X	X	X	X	X	X	X	X	X	X
ASCC, theater	X	X	X	X	X	X	X	X	X	X
Corps	X	X	X	X		X	X	X	X	X
Division	X	X	X	X		X	X	X	X	X
Brigade, BCT	X	X	X	X		X	X	X	X	X
Battalion	X	X	X				X	X	X	X
Company	X	X	X				X	X		X

Echelon/Capability #	11	12	13	14	15	16	17	18	19	20
Joint, combatant command	X	X	X	X	X	X	X	X	X	X
ARFORCYBER	X	X	X	X	X	X	X	X	X	X
ASCC, theater	X	X	X	X	X	X	X	X	X	X
Corps	X	X	X	X	X		X	X	X	X
Division	X	X	X	X	X		X	X	X	X
Brigade, BCT	X	X	X	X	X		X	X	X	X
Battalion	X	X		X			X	X	X	
Company	X	X		X			X	X	X	

Echelon/Capability #	21	22	23	24	25					
Joint, combatant command	X	X	X	X	X					
ARFORCYBER	X	X	X	X	X					

Echelon/Capability #	01	02	03	04	05	06	07	08	09	10
ASCC, theater	X		X	X	X					
Corps		X	X	X	X					
Division		X	X	X	X					
Brigade, BCT		X	X	X	X					
Battalion			X	X	X					
Company			X	X	X					

d. CyberWar. To conduct CyberWar effectively, the future force requires the capabilities as listed in table D-3 below. Access is the first and most critical requirement for CyberWar capabilities and is therefore a major point of emphasis in the required capabilities below. More detailed CyberWar capabilities are included in the classified appendix E.

Table D-3
Required CyberWar capabilities

#	Required CyberWar Capability	DOTMLPF	Source/ References
1	Access targeted networks, systems, or nodes by both remote and direct means in order to ensure required access to enable CyberWar actions on fleeting targets.	D,O,T,L,M,P,F	TRADOC Pam 525-7-8
2	Enable recurring access to targeted networks, systems, or nodes by both remote and direct means in order to ensure required access to enable CyberWar actions.	D,O,T,L,M,P,F	TRADOC Pam 525-7-8
3	Access adversary hardware and software by both remote and direct means in order to ensure required access to enable CyberWar and CyberSpt actions.	D,O,T,L,M,P,F	TRADOC Pam 525-7-8
4	Access, collect, and exploit adversary cyber targeted information by both remote and direct means in order to detect, deter, deny, and defeat adversary actions and freedom of action.	T,M,P	TRADOC Pam 525-7-8
5	Enable the ability to aggregate, manage, decrypt, linguistically translate, analyze, and report on all data collected in knowledge management systems in order to support CyberOps actions and critical battle command and other capabilities.	D,T,M,P,F	TRADOC Pam 525-7-8
6	Provide remote and expeditionary CyberWar capabilities in order to detect, deter, deny, and defeat adversary actions and freedom of action.	T,O,M,P,F	TRADOC Pam 525-7-8
7	Provide automated sensor-enabled network attack and intrusion detection capability in order to detect, deter, deny, and defeat adversary actions, integrate defense-in-depth with CyNetOps, ensure friendly freedom of action, and deny adversary freedom of action at the time and place of our choosing.	T,M,P	TRADOC Pam 525-7-8 UQ 09 SWG
8	Attack (deny, degrade, disrupt, deceive, destroy) adversary	D,O,T,M,L,P	TRADOC

#	Required CyberWar Capability	DOTMLPF	Source/ References
	networks and critical infrastructure in order to detect, deter, deny, and defeat adversary actions and freedom of action.		Pam 525-7-8
9	Provide sensory enabled network attack and intrusion response capability in order to detect, deter, deny, and defeat adversary actions, integrate defense-in-depth with CyNetOps, ensure friendly freedom of action, and deny adversary freedom of action at the time and place of our choosing.	T,M,L,P	UQ 09 UQ 09 SWG TRADOC Pam 525-7-8
10	Attack adversary networks in order to detect, deter, deny, and defeat adversary actions and freedom of action.	T,M,P	QRM
11	Attack (deny, degrade, disrupt, deceive, destroy) cyber embedded processors and controllers of adversary equipment and systems in order to detect, deter, deny, and defeat adversary actions, integrate defense-in-depth with CyNetOps, ensure friendly freedom of action, and deny adversary freedom of action at the time and place of the Army's choosing.	T,M,L,P	QRM
12	Provide situational awareness of adversary and other specified networks in order to increase the commander's overall SA and enable CyberOps and the commander's overall integrated actions.	O,M,P,F	QRM
13	Map and understand adversary and other specified network structures in order to enable all aspects of CyberOps.	D,T,L,P	QRM
14	Track, locate, and predict adversary activities in cyberspace in order to enable CyberSA, CyberWar, and CyNetOps.	D,O,T,L,M,P,F	QRM
15	Attack adversary information in order to dissuade, undermine, and deceive adversaries and support the commander's overall mission objectives.	D,T,L,P	QRM TRADOC Pam 525-7-8
16	Mitigate or bypass adversary cyber defensive measures in order to execute CyberWar capabilities.	O,T,L,M,P	QRM
17	Impact adversary cyber infrastructure in order to support effective CyberOps actions and the commander's overall mission objectives.	D,O,T,L,M,P,F	JCA

e. Table D-4 below lists the echelon at which each capability is required.

Table D-4
Required CyberWar capabilities by echelon

Echelon/Capability Number	1	2	3	4	5	6	7	8	9	10
Joint, combatant command	X	X	X	X	X	X	X	X		X
ARFORCYBER	X	X	X	X	X	X	X	X	X	X
ASCC	X	X	X	X	X	X	X	X	X	X
Corps	X	X		X	X	X	X	X	X	X

Echelon/Capability Number	1	2	3	4	5	6	7	8	9	10
Division	X	X		X	X	X	X	X	X	X
Brigade, BCT	X	X		X	X	X	X	X	X	X
Battalion				X	X			X		X
Company										

Echelon/Capability Number	11	12	13	14	15	16	17
Joint, combatant command	X	X	X	X	X	X	X
ARFORCYBER	X	X	X	X	X	X	X
ASCC	X	X	X	X	X	X	X
Corps	X	X	X	X	X	X	X
Division	X	X	X	X	X	X	X
Brigade, BCT	X	X	X	X	X	X	X
Battalion	X			X			
Company							

f. CyberSpt. To conduct enhance and enable effective CyNetOps and CyberWar activities, the future force requires the capabilities as listed in table D-5 below.

Table D-5
Required CyberSpt capabilities

#	Required CyberSpt Capability	DOTMLPF	Source / References
1	Perform the cyber aspects of site exploitation in order to support all aspects of CyberOps and the commander's overall intelligence and mission objectives.	D,T,M,L,P	UQ 09 SWG
2	Perform exploit reverse engineering and analysis in order to support and enable effective CyNetOps and CyberWar actions.	D,T,M,P	TRADOC Pam 525-7-8
3	Conduct legal and policy analysis of targeting in order to support CyberWar actions and the commander's decision-making processes.	D,T,M,L,P	TRADOC Pam 525-7-8
4	Conduct penetration testing in order to support of vulnerability and threat-based security assessments.	D,O,T,M,L,P,F	TRADOC Pam 525-7-8
5	Conduct threat-based security and vulnerability assessments in order to develop CyberSA and support effective CyNetOps and DCyD.	D,O,T,M,L,P,F	TRADOC Pam 525-7-8
6	Conduct robust RDT&E of cyber capabilities in order to develop and adapt emerging technologies and solutions to adversary technologies and TTPs for effective CyNetOps and CyberWar.	D,O,T,M,L,P,F	TRADOC Pam 525-7-8
7	Project future adversary cyber capabilities in order to develop and adapt emerging technologies and solutions to	D,O,T,M,L,P	TRADOC Pam 525-7-8

	adversary technologies and TTPs for effective CyNetOps and CyberWar.		
8	Identify opportunities to exploit adversary operations in cyberspace in order to support all aspects of CyberOps and the commander's overall intelligence and mission objectives.	D,O,T,M,L,P	TRADOC Pam 525-7-8
9	Conduct cyber forensics in order to support all aspects of CyberOps and develop and adapt emerging technologies and solutions to adversary technologies and TTPs for effective CyNetOps and CyberWar.	D,O,T,M,L,P	TRADOC Pam 525-7-8
10	Conduct mitigation and remediation for cyber intrusions or attacks in order to develop solutions to adversary technologies and TTPs for effective CyNetOps and CyberWar.	D,O,T,M,L,P	TRADOC Pam 525-7-8

g. Table D-6 below lists the echelon at which each capability is required.

Table D-6
Required CyberSpt capabilities by echelon

Echelon/Capability Number	1	2	3	4	5	6	7	8	9	10
Joint, combatant command	X	X	X	X	X	X	X	X	X	X
ARFORCYBER	X	X	X	X	X	X	X	X	X	X
ASCC	X	X	X	X	X	X	X	X	X	X
Corps	X	X	X	X	X	X	X	X	X	X
Division	X	X	X	X	X			X	X	X
Brigade, BCT	X		X					X	X	X
Battalion	X							X	X	
Company	X								X	

h. CyberSA. CyberSA processes, capabilities, and procedures need to be developed in order to properly contribute to the commander's overall SA, missions, and objectives. To develop CyberSA and to support CyNetOps and CyberWar, the future force requires the capabilities as listed in table D-7 below.

Table D-7
Required CyberSA capabilities

#	Required CyberSA Capability	DOTMLFP	Source / References
1	Provide continuously updated, operationally relevant CyberOps information of friendly, adversary, and other specified cyberspace into the COP in order to more fully develop the commander's overall SA and support decision-making processes. **Note:** This capability encompasses and supports the BCEC capability of, "Display / Share Relevant Information: The Army's battle command system must enable the	T,M,P,F	OF 09 TRADOC Pam 525-7-8 BCEC

#	Required CyberSA Capability	DOTMLFP	Source / References
	visualization and dissemination of essential information for display on the COP. This includes symbols, graphic control measures, friendly and enemy information, civil considerations, and the operational environment."		
2	Provide the commander with real time visibility of units' networks and impact to mission objectives in order to develop more fully the commander's overall SA and support battle command and other key capabilities.	O,T,M,P	UQ 09 SWG
3	Provide the commander with real time visibility of adversary actions on his units' networks and impact to mission objectives in order to more fully develop the commander's overall SA and support battle command and other key capabilities.	O,T,M,P	UQ 09 SWG
4	Provide commanders the understanding of the operational significance of decisions made in reference to actions on the network in order to support the commander's decisionmaking processes.	T,L,P	UQ 09 SWG
5	Provide the commander with sufficiently detailed information, including intelligence gain and loss considerations, in order to support the commander's decision-making processes regarding cyberspace and CyberOps.	D,T,M,L,P	UQ 09 SWG
6	Coordinate collection and sharing efforts with sister services, industry, corporate, contractors, government and other partners in order to more fully develop CyberSA, the commander's overall SA, and properly coordinate, synchronize, and integrate overall operations.	D,T,M,L,P	TRADOC Pam 525-7-8 UQ 09 SWG
7	Understand and integrate into the social and cultural aspects of social networking sites and activities in order to more fully develop CyberSA, the commander's overall SA, and the integration of CyberOps into the commander's objectives and missions,.	D,T,M,P	TRADOC Pam 525-7-8
8	Identify potential cyber threats, including potential adversaries in order to develop the commander's SA and support operational and intelligence objectives.	D,T,M,L,P	TRADOC Pam 525-7-8
9	Develop awareness of the attack and defense postures of potential adversaries in order to provide enable effective CyberOps planning, intelligence, and operations.	D,T,M,P	TRADOC Pam 525-7-8
10	Assess the motives, goals, and calculations employed by potential adversaries in their decision to conduct a cyber attack against U.S. or friendly assets in order to enable effective CyberOps planning, intelligence, and operations.	D, T,M,L,P	TRADOC Pam 525-7-8

i. Table D-8 below lists the echelon at which each capability is required.

Table D-8
Required CyberSA capabilities by echelon

Echelon/Capability Number	1	2	3	4	5	6	7	8	9	10
Joint, combatant command	X	X	X	X	X	X	X	X	X	X
ARFORCYBER	X	X	X	X	X	X	X	X	X	X
ASCC	X	X	X	X	X	X	X	X	X	X
Corps	X	X	X	X	X	X	X	X	X	X
Division	X	X	X	X	X	X	X	X	X	X
Brigade, BCT	X	X	X	X	X		X	X	X	X
Battalion	X	X	X	X	X		X			
Company		X		X	X		X			

Appendix E
Required Capabilities Supplemental (Classified)

Classified - Published Separately

Appendix F
DOTMLPF Integrated Question List

F-1. Introduction

a. There are important implications for the services as the Army develops CyberOps capabilities. The integration and synchronization of CyberOps capabilities across the DOTMLPF domains is required in the context of joint interdependence. This integration and synchronization must take advantage of, and be informed by, previous and on-going efforts to include, but not limited to, the joint and Army EW CBAs, the Army Space CBA, the Army EMSO CBA, the Army Network Transport CBA, the Strategic Command Quick-Look Cyber CBA, the Office of Secretary of Defense - Intelligence Computer Network Attack CBA, and the Homeland Defense CBA. These efforts demonstrate significant joint CyberOps interdependence. While some study issues may go beyond the Army's direct role or responsibility, the ability to influence the design and development of the range of DOTMLPF solutions for the joint force as they apply to required capabilities for land operations is an Army responsibility. Specific CyberOps capabilities, such as requirements for Army expeditionary operations, should be examined and detailed for an integrated effort by the joint and Army communities.

b. The Army's family of concepts was used in the development of this pamphlet and each concept includes a discussion of the associated DOTMLPF implications. Several of the concepts identify implications that directly relate to CyberOps and are explicit enough to generate action for change within the DOTMLPF domains. Responsible proponencies and agencies have identified battle command, EW, or other capabilities as many of these implications. These DOTMLPF implications must be identified to avoid unnecessary overlaps and redundancies, to support their integration and synchronization to ensure there are no capability gaps, and leveraged to create asymmetric operational advantages by adapting technologies and capabilities across functional lines. The following concepts should be of particular note, TRADOC Pam 525-5-600, TRADOC Pam 525-7-6, and TRADOC Pam 525-7-16.

F-2. Implications

a. The primary DOTMLPF implications arising from this pamphlet vice an exhaustive list, are described below. The items cited will require additional analysis before comprehensive actionable recommendations emerge. This is intended to be used as one of the starting points for the Army CyberOps CBA.

b. How will the Army concurrently develop CyberOps doctrine with the evolution the cyberspace domain and USCYBERCOM; unfolding joint CyberOps and Army concepts and doctrine; and emerging joint and Army CyberOps requirements?

c. What are the most effective organizational designs for implementation of the Army CyberOps that support USCYBERCOM and all combatant commanders; and support Army expeditionary operations, and an Army enterprise construct that includes the operating and generating forces, active and reserve components, and other specified cyberspace?

d. What is the appropriate balance between joint and Army training for CyberOps personnel?

e. How should the Army work with the joint force in developing robust and responsive CyberOps RDT&E, and acquisition processes?

f. How will the Army keep pace with the rate of technologic advancement, and adapt these advancements to operational advantage?

g. How does the Army develop leaders that understand cyberspace and CyberOps; understand how to integrate and employ CyberOps as part of FSO; and understand how to recognize and adapt advance technologies into operational advantage?

h. How does the Army best organize to develop the requisite CyberOps personnel skills and capacity?

i. What test and training facilities are required to support CyberOps and the integration of CyberOps into FSO?

j. What are the public-private partnerships required to secure commercial segments "inside" the Army or other specified cyberspace?

F-3. Doctrine

a. Emerging joint and Army doctrine must fully capture the description and implications of cyberspace. It must also expand upon the ideas presented in this pamphlet to guide further combat development. The nature of CyberOps will likely require a multilevel security-tiered doctrinal approach. What will make the initial CyberOps doctrine development unique is that joint and Army concepts are still evolving and that joint and Army doctrine will be developed concurrently and likely collaboratively.

b. Doctrine questions include, but are not limited to, the following.

(1) How will the Army concurrently develop CyberOps doctrine with the evolution the cyberspace domain and USCYBERCOM; unfolding joint and Army CyberOps concepts and doctrine; and emerging joint and Army CyberOps requirements?

(2) Does current Army doctrine adequately address Army CyberOps capabilities and integration with the other Army operations and functions and with joint operations throughout a joint campaign?

(3) How does emerging joint CyberOps doctrine influence the conduct of Army operations?

(4) Does joint and Army doctrine adequately address the joint interdependence in the area of cyberspace and CyberOps?

(5) What are the impacts of national and international law on joint and Army CyberOps doctrine?

(6) Are cyberspace and CyberOps adequately addressed in Army doctrine for the theater, corps, and division (and below) doctrinal publications?

(7) Are current TTPs adequate to execute required Army CyberOps operations?

(8) Do proponent doctrinal publications integrate requisite Army CyberOps?

(9) What emerging CyberOps technologies, processes and capabilities need to be codified in Army doctrine?

(10) What types of joint command and control and battle command operations may be impacted by Army CyberOps doctrine?

(11) Does the CyberOps doctrine help the commander on the ground?

(12) Does the CyberOps doctrine provide the commander with an ability to effect operations?

F-4. Organization

a. Army organizations must support USCYBERCOM and other combatant commander requirements as well as the Army's generating and operating forces' needs to include requirements for expeditionary operations, active and reserve components, and other relevant cyberspace.

b. Organizational questions include, but are not limited to, the following.

(1) What are the appropriate organizational structures to enable effective Army CyberOps?

(2) Are current Army organizations adequate to meet the CyberOps requirements of the future force?

(3) Can current organizational structures be augmented to satisfy the capability requirements of Army CyberOps?

(4) Is a new organizational structure required to achieve the required CyberOps capabilities?

(5) What Army CyberOps capabilities should reside at each echelon in Army tactical and operational forces as well as generating forces?

F-5. Training

a. Army CyberOps personnel must be trained to joint and Army standards. There are current CNO and NetOps training venues and programs already established that can be leveraged and modified to meet initial cyber requirements. Training requirements will likely be driven by USCYBERCOM requirements and Army specific requirements necessary to support generating and operating force requirements.

b. Training questions include, but are not limited to the following.

(1) How is the integration and application of cyberspace and CyberOps capabilities included in current training and leader development?

(2) How can the Army adapt its training to better integrate Army CyberOps?

(3) How much training and what training standards are going to be directed from USCYBERCOM?

(4) How will evolving technologies and ongoing or planned changes in organization affect the ways in which Army units and leaders operate and what are the training implications of these changes to support Army CyberOps?

(5) How will evolving CyberOps doctrine impact units and leaders?

(6) What training designs will develop units and leaders able to capitalize on the full range of CyberOps capabilities?

(7) What are the CyberOps training requirements for enlisted personnel, noncommissioned officers, officers, DA civilians, and contractors?

(8) What training and education is required for a CyberOps planner and integrator on a USCYBERCOM and combatant command staff, on generating force staffs, and on operational staffs from ASCCs down to company level?

(9) What type, scope, and frequency of Army CyberOps training must the future force conduct to enable effective operations?

(10) What CyberOps test and training ranges are necessary?

(11) What modeling and simulations are required to support Army CyberOps operations at the tactical, operational and strategic levels?

(12) What joint CyberOps training is necessary and for whom?

(13) What national and commercial agency CyberOps training is necessary and for whom?

(14) What are the impacts of training requirements on the schools (that is, growth, resources, and others)?

(15) How can CyberOps training be conducted at the individual and collective levels?

(16) How will CyberOps be trained and evaluated in units prior to deployment?

F-6. Materiel

a. CyberOps are technical in nature and materiel solutions will need to be rapidly developed, tested, evaluated, and acquired. The utility of materiel solutions will likely be temporal in value due to the rapid technologic advancements and proliferation. Adversary counters to cyber tools also result in combat development being a continuous, fast paced process. Materiel solutions are also expensive and will likely be developed using joint, interagency, and public-private partnerships to take advantage of existing best practices and maximize resource utilization.

b. Some significant materiel questions include, but are not limited to the following.

(1) How will compatibility and interoperability, as well as operational effectiveness be achieved for CyberOps systems?

(2) What technologies are critical to consider and invest in, for the development of effective and capable CyberOps materiel solutions?

(3) Given the crowded nature of the EMS, how can the Army effectively operate in cyberspace and the EMS?

(4) Given the technical nature of CyberOps systems, how will the Army develop personnel and organizations capable of effectively executing CyberOps planning, command and control, operations, and maintenance?

(5) How will the Army effectively integrate CyberOps capabilities given the highly interdependent and interrelated nature of CyberSA, CyNetOps, CyberWar, CyberSpt, EA, EP, and ES systems?

(6) Since CyberOps capabilities often have a short shelf life, what is the most effective and efficient way to use RDT&E and acquisition processes to gain and maintain advantage?

(7) Will TRADOC Pam 525-7-8 unite commanders and technology in such a way as to enable both to successfully meet future non-linear challenges?

F-7. Leadership and education

a. Commanders, staffs, and Soldiers must be educated to understand cyberspace and CyberOps. Doctrine will provide the intellectual foundation to prepare leaders for how to think CyberOps in the context of the future OE. Leader development and education will provide leaders with the understanding of how to integrate CyberOps into FSO.

b. Leader development questions include, but are not limited to, the following.

(1) How does the Army develop adaptive leaders that understand cyberspace and CyberOps; know how to integrate and employ CyberOps as part of FSO; and are able to recognize and adapt advance technologies into operational advantage?

(2) How do we provide collaborative, distributed training aids that support commanders, as well as staffs during planning, preparation, rehearsal, and execution of cyberspace exercises and integration?

(3) How can CyberOps be incorporated into training exercises and leader development to develop CyberOps planning and operations?

(4) How does the Army best educate leaders and Soldiers to understand the complex and ever-changing future OE?

F-8. Personnel

a. The Army must have sufficient trained CyberOps personnel with the requisite knowledge, skills, and attributes. Emerging joint and Army requirements warrant a complete analysis of personnel requirements and the most effective way to develop and manage them.

b. Personnel questions relating to CyberOps include, but are not limited to, the following.

(1) How do we recruit and retain the personnel necessary to perform Army CyberOps functions?

(2) What skill sets are required in Army civilian and contractor support personnel?

(3) What is the best means of selecting Army CyberOps officers?

(4) Should the Army precommissioning programs include a CyberOps component?

(5) What is the right mix of personnel between CyberOps professionals and other personnel selected to serve in CyberOps related positions?

(6) What will be the CyberOps personnel impacts as they relate to other proponents?

(7) What will be the personnel end strength impacts as related to required CyberOps capabilities?

F-9. Facilities

a. This pamphlet anticipates significant investment in the facilities and infrastructure necessary to support CyberOps. The ability to effectively and efficiently conduct test, training, and operations using CyberOps systems will require a robust infrastructure. Such facilities and infrastructure must allow networked and distributed operations as well as multilayered security constructs. The planning and resourcing for facility and infrastructure must be initiated with sufficient lead time to reach maturity synchronous with the future force and anticipated technology developments associated with cyberspace and CyberOps.

b. Facilities questions include, but are not limited to, the following.

(1) Are there adequate facilities available to effectively develop, test, and train CyberOps capabilities so that personnel understand and have confidence in the system?

(2) What infrastructure is required at Army and DOD installations to adequately support Army CyberOps programs consistent with joint, Army, and multinational concepts and as specified joint national training center attributes?

(3) What infrastructure is required in theater to support CyberOps missions?

Glossary
Section I
Abbreviations

ARCIC	Army Capabilities Integration Center
ARFORCYBER	Army Forces United States Cyber Command
ARNG	Army National Guard
ASCC	Army Service Component Command
BCEC	battle command essential capabilities
BCT	brigade combat team
CBA	capabilities based assessment
CAC	Combined Arms Center
CG	commanding general
CIKR	critical infrastructure key resources
CNA	computer network attack
CND	computer network defense
CNE	computer network exploitation
CNO	computer network operations
CONOPS	concept of operations
CONUS	continental United States
COP	common operational picture
COTS	commercial off the shelf
CyA	cyber attack
CyberOps	cyberspace operations
CyberSA	cyber situational awareness
CyberSpt	cyber support
CyberWar	cyber warfare
CyCM	cyber content management
CyD	cyber defense
CyE	cyber exploitation
CyEM	cyber enterprise management
CyNetOps	cyber network operations
DHS	Department of Homeland Security
DA	Department of the Army
DOD	Department of Defense
DOTMLPF	doctrine, organization, training, materiel, leadership and education, personnel, and facilities
DSCA	defense support to civil authorities
DCS	distributed control system
DCyD	dynamic cyber defense
EA	electronic attack
EMS	electromagnetic spectrum
EMSO	electromagnetic spectrum operations
EP	electronic protection
ES	electronic warfare support
EW	electronic warfare

FM	field manual
FSO	full spectrum operations
GIG	global information grid
GNE	global network enterprise
GNEC	global network enterprise construct
GOTS	government off the shelf
ICD	initial capabilities document
ICDT	integrated concept development team
ICT	information and communications technology
IO	information operations
IP	Internet protocol
JCA	joint capability area
JCIDS	Joint Capabilities Integration and Development System
JIIM	joint, interagency, intergovernmental, and multinational
JP	joint publication
JTF	joint task force
MCO	major combat operations
NetOps	network operations
NOSC	network operations and security center
OE	operational environment
OF	Omni Fusion
OV	operational view
Pam	pamphlet
QRM	quadrennial roles and missions report
RCERT	regional computer emergency response team
RDT&E	research, development, test and evaluation
RF	radio frequency
ROE	rules of engagement
SA	situational awareness
SBCT	Stryker brigade combat team
SCADA	supervisory control and data acquisition
SECDEF	Secretary of Defense
SIGINT	signals intelligence
SME	subject matter expert
SWG	seminar wargame
TNOSC	Theater Network Operations and Security Center
TRADOC	Training and Doctrine Command
TSCP	Theater Security Cooperation Plan
TTP	tactics, techniques, and procedures
UQ	Unified Quest
U.S.	United States
USAR	United States Army Reserve
USCYBERCOM	United States Cyber Command
USNORTHCOM	United States Northern Command
USPACOM	United States Pacific Command

Section II
Terms

References for the terms are listed behind the definition. If no reference is listed, Joint Publication (JP) 1-02 is the source of the definition.

battle command
The art and science of understanding, visualizing, describing, directing, leading, and assessing forces to impose the commander's will on a hostile, thinking, and adaptive enemy. Battle command applies leadership to translate decisions into actions—by synchronizing forces and warfighting functions in time, space, and purpose—to accomplish missions (FM 3-0).

combined arm
Combined arms is the synchronized and simultaneous application of the elements of combat power to achieve an effect greater than if each element of combat power was used separately or sequentially (FM 3-0).

computer network attack
Actions taken through the use of computer networks to disrupt, deny, degrade, or destroy information resident in computers and computer networks, or the computers and networks themselves (JP 1-02).

computer network defense
Actions taken to protect, monitor, analyze, detect, and respond to unauthorized activity within the DOD information systems and computer networks (JP 1-02).

computer network defense response actions
Deliberative, authorized defensive measures or activities that protect and defend DOD computer systems and networks under attack or targeted for attack by adversary computer systems and networks. Response actions extend DOD's layered defense-in-depth capabilities and increase DOD's ability to withstand adversary attacks (Assistant SECDEF Memorandum, "Guidance for Computer Network Response Actions").

computer network exploitation
Enabling operations and intelligence collection capabilities conducted through the use of computer networks to gather data from target or adversary automated information systems or networks (JP 1-02).

computer network operations
Comprised of CNA, CND, and related CNE enabling operations (JP 1-02).

counterintelligence
Information gathered and activities conducted to protect against espionage, other intelligence activities, sabotage, or assassinations conducted by or on behalf of foreign governments or elements thereof, foreign organizations, or foreign persons, or international terrorist activities (JP 2-0).

critical infrastructure protection
Actions taken to prevent, remediate, or mitigate the risks resulting from vulnerabilities of critical infrastructure assets. Depending on the risk, these actions could include: changes in tactics, techniques, or procedures; adding redundancy; selection of another asset; isolation or hardening; guarding, and others (JP 3-28).

cyber attack
CyA actions combine CNA with other enabling capabilities (such as, EA, physical attack, and others) to deny or manipulate information and/or infrastructure (TRADOC Pam 525-7-8).

cyber content management
CyCM is the technology, processes, and policy necessary to provide awareness of relevant, accurate information; automated access to newly discovered or recurring information; and timely, efficient, and assured delivery of information in a usable format (TRADOC Pam 525-7-8).

cyber counterintelligence
Measures to identify, penetrate, or neutralize foreign operations that use cyber means as the primary tradecraft methodology, as well as foreign intelligence service collection efforts that use traditional methods to gauge cyber capabilities and intentions (JP 2-01.2).

cyber defense
CyD is actions that combine information assurance, computer network defense (to include response actions), and critical infrastructure protection with enabling capabilities (such as, EP, critical infrastructure support, and others) to prevent, detect, and ultimately respond to an adversaries ability to deny or manipulate information and/or infrastructure. CyD is integrated with the dynamic defensive aspects of CyberWar to provide defense in depth (TRADOC Pam 525-7-8).

cyber enterprise management
CyME is the technology, processes, and policy necessary to effectively operate computers and networks (TRADOC Pam 525-7-8).

cyber exploitation
CyE is actions combining CNE with enabling capabilities (such as, ES, SIGINT, and others) for intelligence collection and other efforts (TRADOC Pam 525-7-8).

cyber network operations
The component of CyberOps that establishes, operates, manages, protects, defends, and provides command and control of the LandWarNet, CIKR, and other specified cyberspace (TRADOC Pam 525-7-8).

cyber situational awareness
The immediate knowledge of friendly, adversary and other relevant information regarding activities in and through cyberspace and the EMS. It is gained from a combination of

intelligence and operational activity in cyberspace, the EMS, and in the other domains, both unilaterally and through collaboration with our unified action and public-private partners (TRADOC Pam 525-7-8).

cyber support
Those supporting activities which are generated and employed to specifically enable CyNetOps and CyberWar. They include vulnerability assessment and operational force-based security assessment and remediation, reverse engineering malware, cyber aspects of site exploitation, counter intelligence and law enforcement, forensics, RDT&E, combat development, and acquisition (TRADOC Pam 525-7-8).

cyberspace
A global domain within the information environment consisting of the interdependent network of information technology infrastructures, including the Internet, telecommunications networks, computer systems, and embedded processors and controllers (JP 1-02).

cyberspace operations
The employment of cyber capabilities where the primary purpose is to achieve objectives in and through cyberspace. Such operations include computer network operations and activities to operate and defend the GIG (JP 1-02).

cyberspace warfare
The component of CyberOps that extends cyber power beyond the defensive boundaries of the GIG to detect, deter, deny, and defeat adversaries. CyberWar capabilities target computer and telecommunication networks and embedded processors and controllers in equipment, systems, and infrastructure. CyberWar uses CyE, CyA, and DCyD in a mutually supporting and supported relationship with CyNetOps and CyberSpt (TRADOC Pam 525-7-8).

dynamic cyber defense
DCyD actions combine policy, intelligence, sensors, and highly automated processes to identify and analyze malicious activity, simultaneously tip and cue and execute preapproved response actions to defeat attacks before they can do harm. DCyD uses the Army defensive principles of security, defense in depth, and maximum use of offensive action to engage cyber threats. Actions include surveillance and reconnaissance to provide early warnings of pending enemy actions. DCyD is integrated with the defensive aspects of CyNetOps to provide defense in depth (TRADOC Pam 525-7-8).

electromagnetic spectrum
The range of frequencies of electromagnetic radiation from zero to infinity. It is divided into 26 alphabetically designated bands (JP 1-02).

electronic attack
Division of electronic warfare involving the use of electromagnetic energy, directed energy, or antiradiation weapons to attack personnel, facilities, or equipment with the intent of degrading, neutralizing, or destroying enemy combat capability and is considered a form of fires (JP 3-13.1).

electronic protection
Division of electronic warfare involving actions taken to protect personnel, facilities, and equipment from any effects of friendly or enemy use of the electromagnetic spectrum that degrade, neutralize, or destroy friendly combat capability (JP 3-13.1).

electronic warfare
Military action involving the use of electromagnetic and directed energy to control the electromagnetic spectrum or to attack the enemy. EW consists of three divisions: EA, EP, and ES (JP 3-13.1).

electronic warfare support
Division of EW involving actions tasked by, or under direct control of, an operational commander to search for, intercept, identify, and locate or localize sources of intentional and unintentional radiated electromagnetic energy for the purpose of immediate threat recognition, targeting, planning, and conduct of future operations (JP 3-13.1).

frequency deconfliction
A systematic management procedure to coordinate the use of the EMS for operations, communications, and intelligence functions. Frequency deconfliction is one element of electromagnetic spectrum management (JP 3-13.1).

frequency management
The requesting, recording, deconfliction of and issuance of authorization to use frequencies (operate electromagnetic spectrum dependent systems) coupled with monitoring and interference resolution processes (JP 6-0).

full spectrum operations
Army forces combine offensive, defensive, and stability or civil support operations simultaneously as part of an interdependent joint force to seize, retain, and exploit the initiative, accepting prudent risk to create opportunities to achieve decisive results. They employ synchronized action—lethal and nonlethal—proportional to the mission and informed by a thorough understanding of all variables of the operational environment. Mission command that conveys intent and an appreciation of all aspects of the situation guides the adaptive use of Army forces (FM 3-0).

global information grid
The globally interconnected, end-to-end set of information capabilities, associated processes, and personnel for collecting, processing, storing, disseminating, and managing information on demand to warfighters, policy makers, and support personnel. The GIG includes owned and leased communications and computing systems and services, software including applications), data, security services, other associated services and National Security Systems (JP 6-0).

information
Facts, data, or instructions in any medium or form. The meaning that a human assigns to data by means of the known conventions used in their representation.

information assurance
Measures that protect and defend information and information systems by ensuring their availability, integrity, authentication, confidentiality, and nonrepudiation. This includes providing for restoration of information systems by incorporating protection, detection, and reaction capabilities (JP 3-13).

information engagement
The integrated employment of public affairs to inform U.S. and friendly audiences; psychological operations, combat camera, U.S. government strategic communication and defense support to public diplomacy, and other means necessary to influence foreign audiences; and, leader and Soldier engagements to support both efforts (FM 3-0).

information environment
The aggregate of individuals, organizations, and systems.

information operations
The integrated employment of the core capabilities of EW, computer network operations, psychological operations, military deception, and operations security, in concert with specified supporting and related capabilities, to influence, disrupt, corrupt, or usurp adversarial human and automated decisionmaking while protecting the same.

intelligence
The product resulting from the collection, processing, integration, evaluation, analysis, and interpretation of available information concerning foreign nations, hostile or potentially hostile forces or elements, or areas of actual or potential operations. The term is also applied to the activity which results in the product and to the organizations engaged in such activity (JP 1-02).

intelligence preparation of the battlespace
An analytical methodology employed to reduce uncertainties concerning the enemy, environment, and terrain for all types of operations. Intelligence preparation of the battlespace builds an extensive database for each potential area in which a unit may be required to operate. The database is then analyzed in detail to determine the impact of the enemy, environment, and terrain on operations and presents it in graphic form. Intelligence preparation of the battlespace is a continuing process.

intelligence, surveillance, and reconnaissance
Activities that synchronize and integrate the planning and operation of sensors, assets, and processing, exploitation, and dissemination systems in direct support of current and future operations (JP 2-01).

Internet
An electronic communications network that connects computer networks and organizational computer facilities around the world (Merriam Webster).

LandWarNet
The Army's contribution to the GIG that consists of all globally interconnected, end-to-end set of U.S. Army information capabilities, associated processes, and personnel for collecting, processing, storing, disseminating, and managing information on demand supporting warfighters, policy makers, and support personnel. It includes all U.S. Army owned and leased) and leveraged DOD and joint communications and computing systems and services, software including applications), data security services, and other associated services. LandWarNet exists to enable the war fight through battle command (TRADOC Pamphlet 525-5-600).

network enterprise center
Provide local (post, camp, base) tenant units with access to the network, network services, communications, and information enterprise services.

network operations
Activities conducted to operate and defend the GIG (JP 6-0).

network service center
A global network operations and service desk functions, information services, and network connectivity through distributed TNOSCs, area processing centers, and regional hub nodes.

operational environment
A composite of the conditions, circumstances, and influences that affect the employment of capabilities and bear on the decisions of the commander (JP 3-0).

signal
As applied to electronics, any transmitted electrical impulse. Operationally, a type of message, the text of which consists of one or more letters, words, characters, signal flags, visual displays, or special sounds with prearranged meaning, and which is conveyed or transmitted by visual, acoustical, or electrical means.

signals intelligence
A category of intelligence comprising either individually or in combination all communications intelligence, electronic intelligence, and foreign instrumentation signals intelligence, however transmitted. Intelligence derived from communications, electronic, and foreign instrumentation signals.

supervisory control and data acquisition
An electronic system that provides for monitoring and controlling systems or processes remotely (Training Manual 5-601).

telecommunications
Any transmission, emission, or reception of signs, signals, writings, images, sounds, or information of any nature by wire, radio, visual, or other electromagnetic systems (1-02).

Title 10, U.S. Code
This title addresses securing U.S. interests by conducting military operations in cyberspace.

Title 18, U.S. Code
The focus is on law enforcement and the principle agency is the Department of Justice. This title addresses crime prevention, apprehension, and prosecution of cyberspace criminals.

Title 32, U.S. Code
The focus is on the first line of defense of the U.S. The principle agencies are the Army and Air Force National Guards. This title addresses the support to the defense of U.S. interests in cyberspace through critical infrastructure protection, domestic consequence management, and other homeland defense-related activities.

Title 40, U.S. Code
The focus is on CIO roles and responsibilities. All Federal department and agencies are responsible. This title establishes and enforces standards for acquisition and security of information technologies.

Title 50, U.S. Code
The focus is on foreign intelligence and counterintelligence activities. The principle agencies are the intelligence agencies aligned under the Office of the Director of National Intelligence. This title addresses intelligence gathering through cyberspace on foreign intentions, operations, and capabilities.

Title 60, U.S. Code
The focus is on homeland security and the principle agency is the DHS. This title addresses the security of U.S. cyberspace.

warfighting function
A group of tasks and systems. (people, organizations, information, and processes) united by a common purpose that commanders use to accomplish missions and training objectives (warfighting functions).